The Order

Ezekiel 10:6-8

THE ORDER

Beau Peterson

Illustrated by
Debbie Pagliughi

For information write to:
P.O. Box 1646
Swansboro, N.C. 28584

Printed by B & B Office Products
Cedar Point, North Carolina 28584

ISBN 0-89962-842-7

Printed in the United States of America.

My appreciation goes to Debbie Pagliughi, who endured many requests for modifications to her beautiful illustrations, and to Bob Cavanaugh, whose insights and encouragement through the years kept the spark alive.

Thanks, also, to the typesetters, proofreaders and printers for their endless patience and assistance in preparing this manuscript.

For Dave Urman...

A promise kept.

Mark 6:46-51

THE ORDER

TABLE OF CONTENTS

THE ORDER

TABLE OF CONTENTS
(continued)

The Order

A theory that the development of Christianity has resulted from a fundamental misunderstanding of an extraterrestrial intervention into the evolution of man in ancient times by alien beings.

Genesis 3:24

FOREWORD

Starmen and Religion

That this book was written not long after man's first landing on the moon is no coincidence. It was the prerequisite, for only today, in the light of our tentative move into space, does the prospect of life on other planets seem to come into focus and approach the realm of reasonableness. Only now may we speculate with a degree of certainty that in this great universe, which operates under the uniform laws of nature, there could be others like us who evolved and ventured into space. In fact, some must have preceded us by thousands of years, as we shall precede others. And it is precisely this skew in the relative levels of development among the stellar races which produced the religion of the ancient Hebrews. How is this so?

Consider that the earth is rich in religious experience. States Schoeps in his *The Religions of Mankind,* "Nowhere on earth have peoples been found who do not possess a religion." (1) From time immemorial man has been a religious being. As man's primitive body was struggling to cope with the rigors of an ever-changing environment, his mind was busy sorting out the mysteries of his world. The forces of nature both frightened and inspired him. He devised every sort of ritual to influence the forces over which he had no control. He appealed to his deceased relatives to watch over and protect him, and he applied every manner of science to evoke or appease the spirits which seemed to abide in all things. He was a vulnerable, often frightened creature in a vast world.

Long before the great civilizations arose from the dust, a more advanced race of beings came to this planet. As they stepped from their spaceships, they found themselves revered as gods. Their ability to descend from the heavens and sophisticated appearance put them in immediate control. Their ships and equipment became sacred. Their knowledge deified them, and when they departed, they left behind at least one new religion - the ancient Hebrew religion.

Today it is clear religions in their assorted forms have existed long before and after such starman visitations, which is why no world order of religions has ever existed. They are *man made* institutions, being as diverse as their makers. The Hebrew religion was no exception.

The Hebrews repeatedly stressed that their God was a *living god* who was greater than all the other gods worshiped at the time. Indeed, their religion was centered around a mission of starmen! These were not weird green monsters as in science fiction movies. They were humans like us on a mission to earth - a mission that became our destiny.

While it was Erich Von Daniken who popularized this approach to understanding our position in the universe, he was by no means the first to suggest such a notion. Why then did Von Daniken's book *Chariots of the Gods* generate such attention?

The answer is both simple and profound, for it was not a question of some mystical or apocalyptic revelation, but a function of the passage of time. It is only because of a basic social and scientific evolution in our culture that such a theory could have aroused such interest - because we can look at ourselves and understand what happened so long ago.

Great advances in computers, lasers, spaceships, biology, archeology, physics - all of this new knowledge - has opened our eyes to the tremendous possibilities around us. Suddenly the pieces to a gigantic puzzle have begun to come together. The resulting picture, if true, promises to alter fundamental beliefs about God and the role of religion. Only now does that greatest of all religions, which speaks so eloquently of the mighty presence and glory of God, give way to yet a *greater* universe into which we are about to enter.

Endnotes

1. Hans-Joachim Schoeps, *The Religions of Mankind,* Doubleday/Anchor Books, New York, 1966, p. 6.

INTRODUCTION

Canst thou by searching find out God?

- Job XI.7

A Parable

There was a young boy—a rather self-conscious, introverted boy—who suffered from the painful adolescent traumas and moral dilemmas which confront us all at some point or another in our lives when he discovered a void that he hadn't felt before. His allegiance was moving away from home and out into the world. It wasn't that he didn't love his family anymore. It was that universal desire to try his wings. It was that time when we begin to assert our independence in every reckless way and question everything as if our youth was an intolerable burden.

His feelings about nearly everything were blown out of proportion by unbalanced hormones. He was known to alternately lash out at the ones he loved or withdraw completely. Moderation was virtually an unknown territory. While his peers were learning how to interact and work together, his interests began to focus on individual activities. Plants, rocks, aquariums and motorcycles occupied most of his free time throughout his youth, a time when others are well on their way to becoming socialized. And he became lonely.

He knew there was a God, for he had always been taught there was one. But he had never felt a particular desire to seek Him out until he began to feel the isolation which he had before so vigorously sought. At that point a personal God became a very comforting idea, and he wanted to meet Him. Many a night thereafter he lay in his bed staring into the void of his darkened room, and he began to pray.

The first prayers were rather mechanical invocations of his childhood: "Our Father, who art in heaven..." and "Now I lay me down to sleep..." But they soon evolved into lengthy one-sided conversations. He began to ask for guidance, for personal improvement and for some response to his prayers. With awe and wonder he contemplated the being he was told guides and strengthens us, who knows all about us even before we tell Him.

Those were the naive years before the bright shiny visions and ideals are tarnished and beaten into conformity by life's tough lessons, a time when all things are possible.

Perhaps at these points in our lifetimes, more than any others, we seek those deeper meanings and assemble the ideas which will solidify into the philosophical base which will guide us for the rest of our lives. Whereas some children seem to grow up in their comfortable little worlds, finding all answers to life's questions in perfect order, others need more substantiation, and *he* was one of those. Perhaps his turbulent, introspective youth made him question more and believe less on face value. But his emphasis on independence and self-reliance, while hindering his social skills, had apparently cultivated another, very positive skill, and that was a heightened sensitivity to the delicate and intricate balance of life which surrounded him.

He continued to search because he was never as comfortable as his peers with the neat little answers he was given in those pleasant conversations with his religion teachers. While others were apparently content with simpler explanations of life's ambiguities, he sought the deeper truths, especially about how a person relates to his God. But answers to his religious questions often seemed to contradict what he had learned in school. For example the biblical version of creation of the world in seven literal days clearly seemed to contradict the secular version of its evolution over the ages as taught by his science teachers.

How could he sort this out? He continued to pray through his dilemma, trying to weigh the evidence as best he could. He asked for guidance and understanding of his doubts as well as for his good fortune and other rewards ostensibly associated with praying and knowing God.

But he needed, as much as anything, some concrete evidence that this entity which he was trying to learn to depend on was in fact *real* before he could commit himself totally and without reservation. He was getting desperate because he was honestly, privately trying to have communion with God but could not get satisfactory results. His innovative approaches, which in retrospect seem comical, were at least sincere.

While in the mountains on a picnic once, he took time out from catching crawdads in the creek to climb the highest hill in the park. It was a steep hill and a great challenge. When he conquered it finally, he threw his arms back and turned slowly in awe of the deep, boundless sky above. It was like a huge temple where a warm summer breeze carried the fragrant aroma of sage past his nose and the expanse made him a little dizzy.

From his place at the top of the world, he thought of God, for he was alone in what seemed like a holy place, and he called out to Him, first quietly, then louder. Finally he screamed at the top of his adolescent voice, "God, I'm here. Please answer me!" His ragged voice was instantly consumed. Turning into the wind, then about, cocking his ear to the sky, he sought a sign. But there came no thundering voice, not even a wind-borne whisper, not a tingle of supernatural feeling or revelation. So he found a suitable rock and planted himself to await a sign. "If you won't talk to me, how can I believe?" he thought as he scanned the horizon. Perhaps God was testing him, trying to instill faith in him. But how could he have faith without some initial proof? Finally, concluding that perhaps this God was just too sophisticated for him, he proceeded to descend the opposite ridge to the mundane world of picnics, beer-guzzling baseball jocks, defiant children and screaming mothers.

There were other boys with questions like his, sort of. Like Rick Clough, the cloud watcher; a romantic, but a thinker and observer who sometimes shared his wonder about such questions. But there were others, like pudgy Greg Davis, a practical kid from middle-class Baptist folks, who, while entertaining his pointed questions in a very diplomatic way for being so young, nevertheless appeared to have no doubts about his God. Steve Catalano, his high school buddy, was sure God's name was Jehovah and had answers for all his questions taken right out of the Bible.

Greg's God was a fearful deity, and sermons in his church carried a message of fire and brimstone consequences for those who stray from the straight and narrow path. The way Steve spoke about it, the wicked would someday be plucked from the earth, which would then become a paradise for those who remained. He found the differences intriguing and sought other points of view, which he found in a Jewish friend, in a Catholic girl he dated in high school, and in listening to some Mormons who visited his house and talked to the family for a couple of days.

With a broadening perspective, he began to ponder the diversity of religious beliefs around the world, the apparent cruelty in a world that allows innocent people to die or be injured, and what happens to us after we die. "If God is perfect, then why does he allow his creation to be so corrupt?" he wondered.

In high school he became aware of politics, of the Vietnam War, of the more global issues and the seemingly impossible dilemmas faced by his country and the world. He became acutely aware of social pecking orders and of relative wealth. He questioned God's apparent selectivity in his issuance of health, freedom, wealth and beauty in a world of pain, poverty and constraint. It seemed that every day he read of innocent blood that had been shed in a crime or an armed conflict. Wealthy nations were seemingly oblivious to poorer nations. Why so many dichotomies and disparities? Many of life's events seemed so random, so discrete, and yet all was not chaos: He could see patterns and visualize a certain symmetry in this order of things.

There was no dramatic revelation, no specific turning point in his life. But gradually questions began to accumulate in his mind about the order of things that people tried to answer with the Bible but never quite succeeded. He was uncomfortable. Like a child restless to take a thing apart to see how it works, he was fidgeting with what he saw but couldn't understand. He had to learn about these things if all religions were not created equal. He had to know which Heaven to seek and which Hell to avoid. Most of his high school years were spent sorting such things out.

He sought answers from many diverse sources, including several different churches. But for the time being, he could only weigh the evidence for he had no rational basis for a belief in an intangible God, and his conscience told him to hold out for something he could relate to his world.

On July 20, 1969, the boy, now becoming a young man, sat glued to his television set as Apollo 11 Astronaut Neil A. Armstrong spoke the immortal words, "One small step for man, one giant leap for mankind" and set the first human foot on an alien moon. This event marked not only a technological milestone, but a leap in the awareness of man as well. Not only had we broken the bounds of the earth physically, but intellectually we were made to understand the awesome potential represented by that short jaunt into space.

Though no longer interested in science fiction, this seemed to be the consummation of all the Ray Bradbury he had read as a younger boy, and it greatly impressed him. A later interest in UFOs led him to read the cult classic *Chariots of the Gods* by Erich Von Daniken, and with that book, the seed of understanding was sowed. Von Daniken's basic premise was that evidence exists today of the visitation of alien spacemen in our ancient history. One of his key points was that the Bible contains passages describing one or a series of those visitations.

But while he was awed by the Apollo mission and amused by the Von Daniken book, he did not make the connection immediately. In fact, he had more pressing personal questions which he thought bore no relationship to such things.

He married at the age of nineteen, just out of high school, and then joined the Marines to live a patriotic life of adventure and honor. It was through his wife Mollie that he was first exposed to the Catholic Church. He was awed by its ornate churches, mystifying dogmas and regal ceremonies. His premarital catechism classes fueled his desire to understand about God and religion. Confession, penitence, the Trinity, communion, the Pope, the Latin mass, purple robes, veiled women, purgatory—it was all so new and interesting! And then there was *The Exorcist.*

What the hell was an "exorcist" anyway, and why did only the Catholics have one? Perhaps no other movie so affected him. He was totally unprepared to encounter such visions of Hell at his precarious point in understanding. He had vivid nightmares for years after. For a while he was scared straight. He began to pray again with renewed vigor and honesty. That Hell seemed so real and so graphic that there simply had to be a Heaven! Questions to the clergy about the nature and purpose of such practices only elicited the same unsatisfactory answers. He was told to pray for answers, and pray he did.

Eventually his prayers again became less of a comfort and more tenuous and testing. "God, if you're there, " he said, "guide me in your ways, and *show me* that I might believe." Again he wanted some feedback, some sign. He was hungry for proof, *any proof*! He didn't care if it was some concrete thing or just some concrete peace of mind. He just needed his questions answered.

Mollie had no similar doubts. Her Catholicism was firmly rooted in her life, and she could not understand his driving skepticism about her religion. She was a pretty woman, slight of build with a fiery disposition and an iron will. Still she had her doubts if not about her religious beliefs, then about her marriage. Eventually with the stresses inherent in being a military wife and the wife of a man with so many questions and doubts, she decided she wanted out, and so did he.

Soon thereafter he received orders to go overseas for a six-month deployment. This was a logical time for everything to go flying apart, and it did so. He drove Mollie home for good, leaving a brand new mobile home to be repossessed, and barely managed to sell his prized Triumph motorcycle at a loss before departing. He was mentally, physically and emotionally exhausted just in time for the deployment.

Paradoxically the regimen of the Corps and his desire to work off his troubles proved to be a very positive personal experience. Instead of constantly worrying about the home front, he began to take an interest in his job as an aircraft electrician. Within a few months after arriving overseas, he managed to get his life pretty well sorted out except for that one nagging facet—his philosophy of life. Up to that point he had always claimed a loose affiliation with one faith or another, but one day he was ordered to the administrative section of his squadron to supply information for his new set of ID tags. When the corporal asked for his religious preference, he had to stop and think. For the first time in his life, he had no preference.

That afternoon during his lunch break, he went to the base book store and bought one of the most important investments he ever made: his Bible. It was a rain-slick, bone-chilling Japanese winter. As he cut through the sleet back to his work space, he felt a new sense of serenity in making the decision to read it. It was a click in his attitude because he was now going to uncover the mystery once and for all.

The squadron's skipper was a hard-charger, a man with that rare ability to reach down to the lowest ranks and inspire his men to follow with enthusiasm. The young Marine admired him and his goal of taking the twelve "pigs" of airplanes the Navy didn't want to fool with anymore and meeting or beating all of the objectives for the deployment.

It seemed like they never stopped. But there were precious moments when one could find time to smoke a cigarette, play cards or read. He was frequently thereafter to be found in some corner of the VMA-224 Electric Shop reading that Bible, for he had made it his goal to read it once and for all, cover to cover. He carefully underlined sections where he had questions, and frequent discussions were generated over the significance of the various passages and stories. He especially valued the constructive criticisms and insights of his bosses, Staff Sergeants Robert Cavanaugh and Jim Dann, regarding his emerging philosophy of religion.

It took more than a year for him to read the Bible from cover to cover, but he could not rest until he had satisfied his mind as to the nature of that elusive God and of that mysterious assemblage of beliefs known as Christianity. From there he could set out to test the scriptures, and for the first time in his life, he could know how to interpret what other people were saying about God. No longer could someone quote a passage out of context and leave him wondering, for he had read the *entire* Bible and could go back and review the surrounding material to verify the intent and meaning. However, out of those readings another, completely unforeseen problem arose. The Bible itself generated as many questions as it answered.

As he began to research various topics from the Bible, he found many new questions arising, questions such as: "What are the J Document, the S Narrative and the P Strand?" "What were the politics involved in the inclusion and exclusion of the various books of the Bible?" "Why does the Bible have stories in which there are two different overlapping versions, one in front of the other?" and so on.

He spent many hours in the base library poring over encyclopedias, texts on religions, Bible narratives, secular histories and scientific journals. He drew on many sources and took many notes. He continued to discuss his interests with his religious friends, his atheist friends, his college professors and with various clergy. His inquiry was as balanced as he could make it. He expanded his reading to include books on fringe subjects - such diverse themes as the black arts, parapsychology, faith, the sociology and psychology of religion, etc., so that he could understand, as best he could, the arguments both for and against Christianity.

Eventually his gut feeling about his search led him more in the direction of a secular approach and away from a religious one. He had always tried to understand the order of things in terms of what he felt provided the best answers to his questions. The whole inquiry into the nature and evolution of the Bible seemed to indicate that the Bible was actually a man-made, if not man-inspired, document. During this time he began to revisit the theory proposed by Erich Von Daniken in *Chariots of the Gods* with a renewed interest. In the wake of *Chariots* came a flood of ancient astronaut books. Interesting rebuttals also appeared soon thereafter, such as *Crash Go the Chariots* and *Some Trust in Chariots.*

Irrationality, emotionalism, sensationalism and even outright inaccuracies marked much of the material, which was hastily compiled to capitalize on the public's insatiable appetite for ancient astronaut theories or the rebuttals to them. Von Daniken became the focal point for the counterattacks, for in his fervent search for evidence of visitations by ancient astronauts he, like others, was often accused of ignoring terrestrial explanations for esoteric phenomena which better fit the circumstances.

But Von Daniken and the flood of authors after him who wrote of extraterrestrial visitations in ancient times had undeniably injected one of the most significant ideas for modern times. Could Von Daniken's assertion that extraterrestrials caused the birth of the ancient Hebrew religion possibly be true? He began to study the idea by reading Biblical passages in the light of Von Daniken's assumptions and soon came to the conclusion that the theory had some merit.

Meanwhile he began to discover during his research that most people simply do not understand their Bibles and the fascinating evolution of Christianity! He began to realize that even *church leaders* appear unaware of some aspects of the Book's history or else they have an unwritten code which says, "You may teach the Word, but of its history and evolution, you may not teach of it."

Realizing that many clergymen attend advanced religious training and are eminently qualified to teach religious subjects from a theological standpoint, he began to speculate on why such knowledge is not readily made known to the church membership. Eventually, however, it became quite clear that Church history is so turbulent and so corrupt that little faith could be mustered by the study of such things. Instead church leaders emphasize current or future matters and on an emotional level. Intangible "spiritual" aspects of life are the focus in accordance with the dictates of their respective denominational theologies. History, if taught at all, is heavily censored and even distorted!

He began to lose faith in organized religion at this point. Such institutions had far too many flaws. "What kind of social condition is it that promotes following so blindly?" he wondered. "Can all of these people who attend church regularly truly be finding peace-of-mind in this myriad of movements, each with its own ways of doing things and its own claims of sole-survivorship in a world they claim is doomed?"

His rejection of Christianity as the path to God didn't turn him away from his investigation, however, because religious movements were still fascinating to study, and they weren't completely devoid of merit. Still, his most important step was yet to come. He needed to synthesize his own logical concept of the order of things before he could completely release his mental bonds to the old belief system. It's called "The First Law of Wing Walking" in the Air Corps: Never let go of one thing until you have a hold on something else.

He took a college course in English literature during his last summer in the Marines and before he left to attend college full time. Much of the course was devoted to Samuel Clemens' book *The Adventures of Huckleberry Finn*.

Clemens, it seems, died without ever completely coming to terms with his God, but he had a profound understanding of human nature and of the condition of the churches of his day. As the young man read the book, he found himself agonizing over many of the same moral and dogmatic questions that Huck did (and in fact, Clemens himself did, for Clemens' writings often reflected his views on mankind.)

Clemens' conclusion was that man is basically a good and ethical being. Some of us are inherently much better than others, some much worse, but most of us fall somewhere in the middle of the spectrum. He believed that most religions, in addition, fail when they try to institutionalize a concept like God, and that a person can be a good and ethical being without the institution and dogmas of the Church. In fact, Clemens found the Church to be as much a culprit in the corruption of man as the system it seeks to overcome. With its hypocritical members and its corrupt leadership, it often shares the blame and responsibility for society's ills.

Perhaps Clemens' message was the final straw. Perhaps it was the cumulative testimony of the Church's turbulent history. Or maybe it was his pragmatic, scientific orientation to life. Whatever the cause, he began to release the idea that the Bible was the word of God, the God he had always searched for. Von Daniken's ideas began to appeal to his sense of logic, and so he pursued them, this time to their logical conclusion.

Eventually a very satisfying scenario began to emerge from his research: Based upon much scientific evidence, he concluded that man is not alone in the universe. Given the universal application of the laws and the elements of nature, we must assume that other planets like the earth have evolved throughout the universe, as well as other human-like beings, who have traveled the galaxies just as we have begun to do! Perhaps, long ago, the earth was visited by a superior order of beings, and they were received as gods.

Much later, when the oral traditions of these visits were reduced to writing, and with the lack of the physical presence of these beings, they were *deified* and became "spiritual." It was in reading the Bible in this light that he began to recognize this order which Von Daniken and others had rediscovered, and he was so impressed with this idea that he set about to write it down in a more comprehensive form.

This parable is my story, of course. I believe I have "found out about God," as stated in Job. There *is* a superior order represented in the Bible, but the images of God as portrayed reveal a being *too human* in temperament, *too* flawed in character to be the perfect entity we envision. Moreover, the text of the Bible proves to be incomplete in places, incoherent in others and often contradictory. Several instances of redundancy, blatant editing errors and omissions caused me to ask how this book has remained unchallenged by the masses for so long.

Can it be the masses are just too lazy to investigate, or do they refuse to see the discrepancies? This refusal to look at the secular facts concerning religion and the Bible points to a condition among Christians I encountered regularly. For example, as I discussed some of the problems I encountered with my Christian friends, I began to realize that our disagreements centered around our perceptions of the issues. The things I accepted as fact, they questioned, and the assumptions they could live with, I could not. In other words, I was talking "apples," but they were talking "oranges."

It became apparent that some people do not require substantiation, verification or proof. Their approach is on a more emotional level, where their psychic and emotional needs are met through the fellowship with their church friends and a simple trust, however unfounded, that what they are taught in church will eventually come to pass. Church dogma becomes their ultimate reality. The world around them may be harsh, unforgiving, even cruel, but they draw their strength from their fellowship pool and their faith in the word of the Bible.

Looking deeper, I eventually came to the conclusion that while some people simply require less proof than others, there was also a more sinister aspect of the Church's operation which suggested a form of psychological coercion through the mental mechanisms of guilt, ignorance, fear and peer pressure. This troubles me because I feel that these tactics are used to maintain membership levels and to secure new members for the survival of the institution and work to the detriment of the individual.

Finally, I felt that I was beginning to understand. I was becoming excited by my findings. I felt that I had scaled another giant mountain and was now looking out over the world in all of its myopic, churning wonder. Yet I didn't feel alone. There were others around me, others who were beginning to see this thing emerging, a *social consciousness* as it were.

In the end the question which remained to be answered was "What then *is* God?" This was the great paradox of my search, however, for the "Starmen," as we shall call them, represented in the Bible benefited from a case of mistaken identity which continues to enhance their image in our eyes today. And yet contained within biblical precepts is the spirit of God, the God we always thought we were worshipping.

As an advanced culture, the Starmen surely understood their relationship with the natural order of things better than we. Our study of the Bible still holds a wealth of knowledge and wisdom fit for today and tomorrow. We have only to understand from whence it really came, and how it should be treated, and then we can use it wisely to advance the cause of humanity in the way it was most probably always intended.

So what has the Starman Intervention Theory done with the Lord? It has simply placed Him above the physical earth, stars and universe, back where He belongs in the spirit of goodness and life. He is therefore the machinery of life, the merciless force of nature which seems both unpredictable and arbitrary in many ways. The same force which animates the delicate cloud orders the seasons with impeccable regularity. He is, moreover, the indomitable spirit which resides within the temple of the human body. But while He is represented in the goodness of Jesus, His powers encompass the evil of Satan as well. He is the light *and* the dark, the positive and the negative, the Alpha and the Omega, for He *is* The Order.

The Pillared Angel by Dürer illustrating Revelation 10:1-5

Chapter I
THE TRADITIONAL VIEW

I, for one, will take my understanding prisoner and bring it into obedience to God.

- Martin Luther (1483-1546)

In our modern world, generally thought to be dominated by an ultra-rational, scientific mentality, there remains a great island of beliefs and traditions rooted in a supernatural realm we call Christianity, to which millons of people pledge their allegiance. Indeed this ancient system of values and beliefs forms the very foundation of the American Way, its fiber interwoven into law, custom and culture. Our entire system is built around the worship of a spiritual, that is, intangible force that is today as sophisticated as the minds of men could conceive over the aeons.

Because God is described as an invisible force, every power has been attributed to Him, and every conceivable earthly occurrence has been credited to his authority and design. In one fundamentalist Christian view, God takes an active hand in the most minute of occurrences, as if some great paternal being were directing pawns on an earthly chessboard. But the level of God's participation in man's affairs seems dependent upon whom you ask. To some Christians, God takes an active hand in world affairs. To others, He is simply a private, personal counselor. The fact is, it does not appear certain just who God is or how He works. Each church has its own official view, and each person therein has accepted or rejected church dogma according to his or her own personal formula.

Traditionally God created the universe: the stars, the solar systems, and the Earth, complete with its oceans, plants and animals. Everything, it is said, was created in six literal days, and on the seventh day, God rested. Then God created man in His image - a perfect and supreme ruler over every earthly thing. But man, though designated "ruler of the earth," was placed here to be tested.

This test was to be one of loyalty, for God also created angels, both good and bad (some had to be imperfect to want to defect), who were to minister to man and cause him to make a moral choice - between following God or following the forces of evil under Satan. This spiritual war over the conscience of man is still being waged *despite God's knowledge* of the outcome. For, according to tradition, all men will sin, though some will repent of their evil ways, and God knows, even before it comes to pass, just who those righteous people will be!

From the original "fall" of Adam and Eve, when they took the fruit of the Tree of Knowledge from the Serpent, man has continued to "sin" throughout his existence. Once the entire race of man (it is claimed) was wiped from the face of the Earth in a single great deluge except for one righteous family from whom the entire population of the earth sprang again. There were other times, as well, when man was so rebellious against God's plan for him that he was threatened with destruction, including an incident during the Exodus, when the Lord threatened to annihilate all of the tribe of Israel but relented after an emotional plea by Moses to spare them.

Eventually God sent a flesh-and-blood son, who was actually *also* God, made flesh, with the *full knowledge* that he would be rejected and killed at the hands of the ones whom he was trying to save from their sins. This act was the ultimate sacrifice for man, and any who could accept the Son as the redeemer of mankind from its sins will someday be admitted into an eternal kingdom of paradise as a reward. This paradise will then house the elect few forever in bliss and harmony while the unrepentant "sinners" are to be condemned to an eternity of torture in a fiery Hell beyond all imagination along with all of the fallen angels who precipitated the fall of man in the first place.

This is, of course, a very basic sketch of the fundamentalist concept that every word of the Bible be taken literally. A major reason for the plurality of the Church in America, however, is the controversy over just how the Bible should be interpreted. Therefore the preceding sketch cannot be applied to any particular denomination. But it is clear that no two church organizations are in complete accord over the appropriate organization, manifestation or teachings of the Christian religion. This being the case, let us briefly consider the condition of the Christian mentality.

While the Church institution *has* provided benchmarks by which a person might develop and measure his moral character, it has also claimed to hold the keys to some absolute "truth" of God. Such claims, unfortunately, are not always critically evaluated, raising questions about how the Church works to influence people, and about how people are conditioned to believe what they do.

Regardless of the vehicle chosen to discover God, there is, it seems, some deep-seated need in man not only for a greater paternal spiritual author to lift the burden of ultimate responsibility for his life from his shoulders, but also to help fill in the gaps in his understanding of his baffling world. Man, being a social animal, seems therefore to find comfort in the fellowship of a church, for churches represent primary social pools of people who share common values. The social pools, in turn, provide reinforcement to an individual that he has made the correct choices concerning his beliefs.

In addition, religion may provide a sort of adjustment mechanism to the disparities of life. We all must die, for example. Religion helps reconcile people to their fates. While many people strive to achieve the highest levels of success, many fall short, and so religion may help soothe and adjust people's desires with reality.

Within this context each denomination's members consider themselves the privileged elect, the "tended flock" of the Lord, and they generally give themselves freely to an appointed minister who is supposed to lead them on the proper path to God. The great order they see around them implies some author to create it, and authority requires subordination.

But separating the belief system from the institution sheds some interesting light on the order of things. The Church, as a dynamic, bureaucratic institution, has actually led people far astray from the original message of the Bible. While it has preserved the Bible itself from too much corruption (until recently), its manipulative nature, with underlying agendas often completely unrelated to the religious education of its members, tends to suppress and confound understanding rather than aid it. By examining some of the Church's dogma and manipulative mechanisms, a clearer understanding will emerge on how the message in the Bible has been affected.

Chapter II
THE DILEMMA WITH THE CHURCH

The master of superstition is the people; and in all superstition wise men follow fools; and arguments are fitted to practice, in a reversed order.

- Francis Bacon

One of my first observations about religions during my formative years was the diversity. "Why are there so many different denominations of churches?" I wondered. There are three main divisions of churches in the U.S.: the Catholics, the Protestants and the Jews. Roughly 60% of the U.S. population can be accounted for by the Catholics, the Baptists and the Methodists alone. According to Kenkel and Voland (2), by adding the Lutherans, Presbyterians, Episcopalians and Jews, we can account for 80% of the population. As of 1976, there were nearly 250 Protestant denominations alone, which gives some idea of the diversity of religious identification today. In fact, religious freedom in America has spawned a kind of religious *chaos.* States Kenkel:

The diversity within broad denominations, the numerous small religious bodies, and the seemingly endless formation of splinter groups and new sects are all part of the denominational pluralism that is a prominent feature of religion in the United States. (3)

To the casual observer, this mass of discriminate movements appears somewhat uncertain as to the best method, or set of standards by which mankind should be guided in its dealings with God. All are obviously concerned about the same entity, for they all use the same source book - the Bible, but they agree on little else. Even the content and formats of the bibles used by the various denominations are different. For example, the Protestants may use a *King James* version of the Bible or a *Good News* version with modern wording. The Jews accept the Old Testament, having nothing to do with the New Testament or the Apocrypha and relying mostly on the first five books of the Old Testament, which they call the *Torah*, while the Catholics include all of the Old and New Testaments as well as selected Apocryphal writings. The Jehovah's Witnesses use Bibles with their own wording, and the Mormons use the *Book of Mormon* to supplement the word of the Bible. Sociologically, however, this diversity is the understandable result of centuries of theological evolution with attendant political and social pressures.

Churches are bureaucratic bodies like any other large organizations, and they must be able to relate to their members, for the members are *the lifeblood* of the organization. Without members, there can be no organization. Thus the fundamental goal of the Church is to serve the needs of its *constituents* and to foster a growth in membership. Only then can the organization carry on its work for God. When conflicts between Church dogma and social change occur, such as the equal rights, birth control, divorce, homosexuality and abortion issues, ways *must* be found to reconcile the needs of the members within the context of Church dogma, or the denomination's membership will suffer. New churches were born in this very fashion. Christianity was launched when the Jews refused to accept that Christ was the literal Son of God. The Protestants emerged in various forms during the Reformation of the Sixteenth Century as a result of many disputes, including the issue of which authority should be supreme, the Pope or the Bible.

But why would God allow man's belief system to splinter into rival factions? If his Son came to unite men under a specific set of standards, then how are we to understand this diversity? How are we to choose the correct religion to worship Him? Christian evangelical movements have responded to this question by saying, in effect, that it does not matter which denomination one is affiliated with because the original church was the *invisible* or *universal* church which binds all Christians under one faith. This downplaying of the organization and fostering of the concept of a universal church under fundamental "spiritual" goals seems to be gaining acceptance and momentum. At a time when the larger denominations have been accused of failing to respond to rapid social changes and to accommodate the religious needs of their members, and with the traditional suspicion of large bureaucratic bodies in general, the evangelical movements are seeing a spectacular rise in interest, especially as a result of the new media preachers.

Paradoxically, however, the rise of the evangelical movements appears to be producing even more powerful, gigantic multi-national bureaucracies with awesome influence on social and political matters! Jerry Sholes comments:

With the advent of electronic religion ... the church has been artificially extended into our homes. The offering plate that gets passed by televised religion has grown so large that it includes the contributions of millions of dollars by millions of people. When you endeavor to trace the "good" that those contributions do, you begin to run into muddy water. A local church does good in its community. A large television ministry tends to localize its funds in and around whatever project the television minister wants to "push." No real good ever finds its way back to the communities of those millions who contribute to large television ministries.

What does happen *is that television ministers enjoy extremely high standards of living ... million dollar homes, expensive cars, big business deals, computerized mailing lists, and cash flow patterns that boggle the mind! You begin to suspect that the entire pattern of large electronic ministries is much more closely related to the ego of the television minister than to anything else. (4)*

In many ways the evangelical movements are no different than the churches they sought to break away from. In some ways they are worse. Recent media headlines about Oral Roberts' threat by God to "call him home" if he didn't receive a certain amount of money for a church project, Jimmy Swaggart's highly publicized secret liaisons with a prostitute, and Jim and Tammy Bakker's months-long media circus involving Jim's sexual affairs, bribes, extravagant lifestyle and general mismanagement of church funds have brought that point graphically home.

A church, by its nature, requires subordination to a leader to guide its members to the proper path to God. Most people, for lack of understanding, ability or resolve are unable to be leaders. Therefore when one dynamic individual aspires to the leadership position, he assumes an impressive degree of power and control over the people who choose to follow. The path that leader follows determines the whole character of the church body.

The problem arises when some leader takes a stand contrary to established tradition and then institutionalizes that into a new church. Now there are no longer branches of one church, but two different churches. Multiply this by the number of districts, countries and regions, and let it ferment for centuries, and you have the diversity we see about us today. Charles Taze Russel's *Jehovah's Witnesses*, Mary Baker Eddy's *Christian Scientists*, Joseph Smith's *Mormons*, and Joseph Bates' *Seventh Day*

Adventists, all of which started during the American revival period from the early to mid eighteen hundreds, are examples.

It matters little whether the views held and preached by these leaders are correct or even rational because for that segment of the population which chose to follow the new, dynamic leader, their concept of reality has been modified. The new leader has appealed to some note of accord in each of those individuals. He has won their allegiance, and whether his ideas are right or wrong, the followers will not even critically evaluate them. Their value systems are changed to accommodate the new order, and it was the leader who redefined it for them. Often the followers will identify with the leader despite the fact that his philosophy may be so far out in left field that they'd need binoculars to see it, for the charismatic leader appeals to the *emotions* and downplays the role of intellect in his followers. Creative thinking is frowned upon, except by church leaders. It is a difficult, sometimes dangerous problem, but the practice of subordinating one's rational evaluation to such modes as feelings, intuition or impressions is a key factor in the belief systems of most religions.

What happens when contemporary scientific and religious views conflict? Children, for example, are usually taught only the theory of evolution in public schools. When they attend church, they are taught "creationism," the idea that God created the universe in six literal days. In our generally pragmatic scientific society, such conflicts must inevitably arouse agonizing questions among the young who are seeking to formalize their concepts of life. But the church has a very potent mechanism for developing and maintaining allegiance to its views; it is a form of psychological "bondage."

Few devout Christians will question in earnest any church dogma, doctrine or practice which offends or confuses them. When they do question such things, they will almost never seek an answer outside of their church or the Bible. This psychological condition which has accompanied Christianity almost since its birth is manifest in a fear of losing or being separated from God - something we shall call the "fear factor." This fear would not exist were it not for the gruesome pictures of Hell which have been painted down through history by creative theologians or their allies. Particularly vivid descriptions were contributed by such authors as Dante in his *Divine Comedy* and Milton's *Paradise Lost*.

Hell was an ingenious invention for without Hell, separation from God might merely mean an existence after life similar to the one we have now, perhaps a hope for reincarnation or even simply non-existence. But Christians aren't usually given the choice of some middle ground (except in the former Catholic concept of Limbo for unbaptized children, which has now been abolished). The only institutionalized alternatives are Heaven or Hell. (Even the Catholic Purgatory is just a kind of temporary Hell leading to the eventual repose in Heaven for those who meet the requisite conditions.) If one refuses Heaven (usually for failure to accept Christ as savior), then the only alternative is Hell, for there is no in-between and never has been since the time they were invented at about the time of Jesus.

For those who were never raised with this "fear factor," and for those who accumulated so many unanswered questions that they found the courage to go beyond the Church doctrines for answers, there is a wealth of secular knowledge available which both amplifies the information in the Bible and also puts it into its proper perspective. For our purposes, the interpretation of biblical text from a secular, historical standpoint provides the tools to free the secrets locked in verse for so long. This approach is in direct contrast with the Church teachings, which emphasize the spiritual, emotional and practical use of the Bible. Perhaps ironically, it is the Church's emphasis which may have actually preserved the message long enough for us to develop the technology to understand what the story was *really* about!

In the next chapter we shall discuss some of the concepts held by the Church and try to begin to show why those concepts are defective. Some concepts arose for political reasons, some for lack of knowledge or understanding, others for theological or dogmatic reasons. They will be explored briefly here in the hope that the reader will seek further information. The primary purpose for examining these concepts is to orient the reader to the original "religion" of the Israelites and to gain some insight into how they developed the concept in light of the Starman Intervention Theory.

Endnotes

2. William F. Kenkel and Ellen Voland, *Society in Action,* Canfield Press, San Francisco, 1975, p. 256.

3. *Ibid.*

4. Jerry Sholes, *Give Me That Prime-time Religion,* Hawthorn Books, New York, 1979, p. IX.

Chapter III
WHY TRADITIONAL CONCEPTS ARE DEFECTIVE

...We do not know a truth without knowing its cause.
- Aristotle

Birth of Christianity

One of the most interesting things about the strength and far-reaching effects of Christianity is that it was born not from among Rome's renowned noblemen, nor from among members of the Greek aristocracy, nor even from among the most prestigious leaders that the Jews had to offer, but out of the rubble—the weak, humble masses of deprived and poor. It came to life through the insight of a Pharisee Jew named Saul of Tarsis, who at first persecuted Jesus and then saw utility in his teachings and espoused them as his own. He also saw an unfulfilled need in those humble masses and set about to fill it. For his work he was christened Paul and made a saint.

The fact that Paul had the brilliant insight to captilize on Jesus' teachings and to spread them to *all* men, especially the poor and the poor in spirit, is the reason the former Jewish cult became the dominant religion of some of the world's most powerful nations. Yet the impact of Paul's ideas upon the birth of Christianity is seldom addressed or even acknowledged by the Church. Because the focus is on Jesus, the credit for Paul's contribution to the new theology had to be subordinated to protect Christ's role. But it is clear to many historians that Paul is at least as pivotal to the birth of Christianity as is Jesus, whose teachings only formed the basis upon which the new religion was built.

One reason Paul's ideas were so obviously successful was that they eliminated classes among people in the eyes of God, making Greeks or Romans equal to Jews. Even women had an equal place in the new scheme of things. This radical new theology which Peter and Paul spread around the Middle East on their ministries also provided the masses with something significant, even when their hands were empty—a promise of a glorious new, eternal life after death in paradise, a reward for those who remained faithful to the new teachings.

No longer would the present life be considered of such consequence or burden to a Christian for with death and martyrdom would come the peace and glory of another, more beautiful life in heaven. Toward this end the masses were forged into a powerful political force which all rulers thereafter had to contend with. Constantine the Great recognized this fact when, barely three centuries later in 313 A.D., he made Christianity the official religion of the Roman Empire, the empire that once so ruthlessly oppressed the new religious movement.

Jesus

A second thing to note about the birth of Christianity is that it arose by default when Jesus was rejected by his own people. Jesus came to try to heal the wounds of the Hebrews because of their on-going transgressions against their Lord, and not for all mankind, as is commonly assumed. That he came to minister to his own people is plainly stated: *"I was sent only to the lost sheep of the house of Israel."* (Matt. 15:24) In fact the entire Old Testament was written about this race of Hebrews as God's *only* chosen people. They were a very select group of whom God was extremely jealous and for whom He decimated entire armies that they might establish a place for themselves in Canaan.

Paradoxically, Jesus was crucified because of the catastrophic splits among various sects of his people over just what type of saviour the Messiah was supposed to be. It was at first only a Jewish matter. Some Jews interpreted the scriptures to mean that the coming saviour would be a great warrior in the traditions of Joshua and King David. Others saw him as a gentle lamb who would win enemies over through his heart. When he tried to unite them again, he was rejected as the awaited saviour, labeled an impostor and a trouble-maker, and then killed for fear that he might unite the Jews serving under Caesar and cause too much trouble in an already tumultuous Roman Empire. Then Paul took up Jesus' cause and propagated it to the Gentiles.

The Hebrews

Of ambiguous origins among the Semitic peoples of the fertile crescent, the ancient Hebrews emerged to become the race God chose as his instrument on the earth. But why was this tiny race so special that God concerned himself solely with their welfare, even to the point of destroying other armies so that the Hebrews might establish themselves in a land of their own? The Bible clearly indicates that God's preoccupation with these people was central to his plan for the earth.

He watched over the Israelites like a parent and, indeed, identified Himself as the *direct parent* of this unique people. Because this special relationship is so clearly identified and reaffirmed throughout the Old Testament, it is apparent that the Hebrews were actually "created" separately from the rest of the Earth's peoples, and their mysterious origins become the clear result of this special creation.

Corroborating this idea is the concept of the "Pure Vine" metaphor which runs through the Old Testament and into the Gospels. Stressing the purity of the race, both racially and culturally, it is the Hebrews' repeated corruption of their pure lineage or culture which causes them no end of trouble with their God. The Hebrews were a pure race under the constant scrutiny of a jealous God on a mission. It was *the Hebrews* who were created "in the image of God," and not the whole world, as is taught today.

It is also important to point out that events concerning the Hebrews in the Old Testament occurred exclusively in the region of the Middle East even while other races and cultures were flourishing around the world. The Old Testament deals with the ancient Hebrews and their neighbors in the region, *not* the whole world, as is commonly taught. Archaelogical and anthropological histories amply confirm this. To believe this is so is to understand why all the rest of the world doesn't espouse the teachings of Jesus as their own. It was a localized event in time and space that was subsequently made universal. The other half of the equation is the Old Testament personification of God.

God

When we think of the Christian God of the Bible, we typically think of His New Testament characterization: the merciful, kind, all-knowing spirit God taught by mainstream churches to be the God of the universe. But approaching the Bible stories from the perspective of the Starman Intervention Theory reveals the curious evolution of the concept of God in the Bible.

The form we encounter when we open to the first book of Genesis is the ethereal form of God which is said to have created the universe. It is the force, we read, which established and directs the laws of nature, which gave us stars and Earth and life. Beginning with the creation of man by God, a new picture of God emerges which has a distinct evolution all its own, both in character and in manifestation. The difference in temperament, for example, between the God of the Old Testament and the God of the New Testament are like day and night - two different beings. The wrathful, aloof, disciplinarian God of Exodus, Deuteronomy and Numbers cannot be easily equated to the kind, mysterious, almost fatherly image presented in the early Genesis accounts preceding it. These two characterizations contrast still more dramatically with the benevolent, gentle, almost benign God of later Christianity.

But there is a more curious evolution in God's form. He is presented as a flesh and blood being in his first dealings with man. We are told in Genesis 3:8-11 that He *walked* in the Garden of Eden and talked directly to Adam and Eve. Long after Genesis, however, God is said to have covered himself in clouds, as during the Exodus, when He spoke face to face only to appointed representatives such as Moses and separated himself from direct contact with the masses. And then, finally, He makes his total disappearance after the fall of Jerusalem and is not seen again until after the arrival of Christ. After Christ, He is never again reported seen in material form. Why this evolution?

The answer has only recently become comprehensible, and it will be shown that this metamorphosis was a logical sequence in a chain of events which continues to be misunderstood today. One only has to read the book of Exodus to be struck by the feeling that the tribe of Israel and Moses were under the influence of a superior order, but a distinctly *human-like* order with God's manifestation and temperament appearing more like that of an indignant general than the creator of the universe. He is portrayed as a mighty *"man of war,"* a *"terrible"* and frightening entity, who worked in a very overt, physical manner and not through some spiritual or metaphysical fashion to influence his people. In fact He is several times brought to the brink of totally annihilating the entire race of Hebrews as He did in the great flood but is moved to mercy by both Abraham and then Moses when God reveals his sentiments to them. These sentiments, however, reflect human attitudes and emotions, displayed in frustration and anger, and not those of some supreme and invisible loving force, which we later devised.

Today we typically envision God as merciful, kind and loving. His Old Testament personification is almost forgotten. In fact the New Testament is often *the only* scripture Christians carry in their witnessing to others. He is a more complex and personal God, too, as evidenced by the concept of the Trinity. God is seen by many branches of Christianity as a triple entity in one being. The three forms, Father, Son and Holy Spirit, are one and the same, all at once pre-existent and eternal. In this way Christian thought conceives the maker, the saviour and the minister as separate parts of the same whole. But this is not how God was understood in ancient biblical times.

The trinity concept became an official Church doctrine in 325 A.D. at the Council of Nicaea, 300 years after the crucifixion of Christ. Jesus himself made it clear that he was a separate and distinct person from the God who sent him and never taught such an idea. Some sources trace the origins of this concept of the equality of the three forms back to the first century A.D., to the Jewish philosopher Philo, whose subsequent followers, drawing on pagan examples of triune gods, imported it into the early Christian theology.

The rationale behind the concept is reasonable. As time passed, it was felt that the image of an all-powerful, often vengeful deity tended to distance God from his people. His Old Testament personification was too frightening and too aloof, so Jesus was made equal to God in every way so that God became both flesh and spirit. The effect was to make God seem more approachable for a more personal relationship.

The "Holy Spirit" is often identified with God's presence. A major factor for justifying its inclusion into the concept of the Trinity is said to be its use by Jesus in his baptismal formula to his disciples at Galilee, where he blessed them *"...in the name of the Father, the Son and the Holy Spirit... ."* (Matt. 28:19) After Christ's crucifixion, it was perceived that the Holy Spirit was the personal guiding force of God which works within the individual to minister and console him and was therefore perceived to be another dimension of the same tri-part manifestation of God.

In the Old Testament, however, God clearly already existed in a physical form, as when the Lord came to visit Abraham by the Oaks of Mamre just prior to the destruction of Sodom and Gomorrah, (Gen. 18) and at many other times to be explored later. During this time God was understood as *"The Lord,"* leader of many lesser ranking aides, sometimes identified as *"angels,"* who operated in a quite military fashion to accomplish various missions. Moreover, the complex force we now call the "Holy Spirit" once indicated an entirely different phenomenon during the times of Moses, Joshua and David. It was not a personal guiding force, but a deadly phenomenon akin to electricity or radiation! This "spirit" was associated with the presence of the Starmen in an operational sense, not in a spiritual sense, and was seen to lift people and objects physically into the air. Improperly handled, it was even known to *kill*.

The power of this "spirit" is made quite clear in passages describing how the priest Uzzah was unmercifully struck dead when he improperly touched the Ark of the Covenant or how Ezekiel the priest in exile was lifted and carried through the air by this same "spirit." Jesus experienced the spirit of the Lord descending "like a dove" from above him before being led by it into the wilderness to commence his preparation for his ministry, and in II Kings 2:16 we read that the "spirit of the Lord" caught up Elijah and took him "into Heaven."

But these forces were understood, even at the times they were witnessed, to be the powers used by God, and not God himself. As time passed, this unseen force became not a tool of God, but a *ghost* of God, a private, personal counselor and a manifestation completely different from what was recorded in the Old Testament.

Satan

Today seen as the force opposing all that is good, the tempter of man, Satan has evolved to become the chief adversary of God in an existence which God supposedly created as perfect. (Deut. 32:4) Contemporary Christians see Satan as the leader of the one-third of the heavenly angels who turned against God sometime before the creation of man with the aim of becoming God's equal. It is said that this rebellion continues today with Satan's fallen angels trying to divert man from the path to God.

Often called demons or devils, these satanic aides are credited with every negative occurrence, every ugly aspect of man's imagination and fault. Satan is, today, as complex and terrifying as nearly 3,000 years of evolution in men's minds could make him. Indeed, his status and sophistication are rivaled only by God, his creator. But here again his form today in no way resembles his form in the Old Testament or even in Jesus' time.

The word *satan* means "adversary" in Hebrew. The Old Testament scarcely mentions this being, however, for early Hebrew references to "adversary" denote *human* opponents. While one element opposing God directly is portrayed in Genesis as the serpent who tempts Eve in the Garden of Eden, little mention occurs again in this context until much later. It is generally believed that the elaborate angelologies and demonologies of the Persians and other surrounding cultures gradually influenced Jewish thought to the point that they became widely accepted and were then incorporated into the inter-testamental literature by the first or second centuries B.C. Demons and Satan thus became more than human adversaries; they became spiritual tempters of man. According to *Harper's Bible Dictionary*, it is in I Chronicles 21:1 that the word *Satan* is first employed as a proper name:

> *It is only in this place, in one of the latest books of the O.T., that the word is used with this particular nuance. From this time on, Satan and other demonic figures begin to appear in Jewish thought; Satan now becomes God's "adversary," not merely men's, and is held responsible for sin and evil. (5)*

Satan's demons evolved from various Assyrian, Babylonian and Greek versions, which have roots in the dim history of man's belief about good and evil spirits. Demons or devils were eventually seen to be the cause of every human malady and became the legion which followed Satan in his mutiny. By the time the New Testament was being written, the grand scheme which we know today had been worked out: God created the universe in its perfect beauty, and He created angels to help him manage his creation. But the most powerful of the angels, mistakenly identified as Lucifer, (6) led a rebellion against God in a power struggle, taking with him perhaps a third of the angels of Heaven. For this act of defiance, God banished the rebellious ones to the Earth, where they began to try to corrupt man and lead him away from God. Thus our time on earth became a test, and this is seen to be continuing today, until the time when God will send these "fallen angels" to their place in Hell and rescue the righteous people who remained faithful to God.

Interestingly, this scenario may have elements which were drawn from a quite different source. One book not included in the Bible, the *Book of Enoch*, argues that there were once angels called "Watchers" who were supposed to monitor man on earth. But a group of Watchers came down to earth against higher authority, mated with the daughters of men, whom they found quite fair, and begat children. For this corruption of the earthmen, the Watchers were banished to *the Earth*, and it was some of the *angels*, not Satan, who precipitated the fall of man. Today only a fragment of this text has survived the editors of our Bibles in Genesis 6:1-4, which describes how the "sons of God" found the daughters of men fair and took wives unto them as they saw fit. From this union were born the "mighty men of old, the men of renown."

What appears to be a difficult scenario to comprehend in the traditional way yields its secrets plainly under the Starman Intervention Theory, as will be demonstrated in subsequent chapters.

Heaven and Hell

A basic Christian tenet today envisions Heaven as the abode of God and his angels, a kind of Utopia or paradise where those who are "saved" from their sins on earth by accepting Jesus as their saviour will be rewarded with an everlasting life after this earthly life. This realm is today largely considered to be a state invisible to our perceptions on Earth. Nevertheless, it is considered a real place, where we will take on new, perfect bodies after our old ones pass away.

Should anyone desire to reject this reward, there awaits a Hell beyond all imagination. This fearful place has been depicted as a constantly burning inferno where souls thirst but are never quenched, where they are tortured and tormented by Satan and his demons and will suffer eternally in damnation. Various artistic renditions have shown people who never shed their decaying bodies and inhabit a subterranean underworld of darkness with lakes of fire in pain and mental anguish. Those who lead lives of sin and do not repent or accept Jesus before they die are thought to end up in Hell.

It is, in fact, the institutionalization of fear for the purpose of reining-in the congregation! Each church denomination may have its own formula, and so the criteria for sinning and being saved may vary from church to church, but the appearance is that Heaven was designed to be the "carrot" and Hell, the "stick." But here again these images of Heaven and Hell in no way represent the original system.

Hell, for example, in its most primitive form, was borrowed from the Greek notion of a dark and dreary underworld called "Hades," or "Sheol" in Hebrew, where the spirits of the dead were thought to dwell after life on the earth. Jesus several times spoke of "Ghenna" or "Ge-Hinnom" (Valley of Hinnom) in referring to what it would be like in such a place. According to Harper's, this gorge had for centuries borne an evil reputation and eventually became the place where the refuse of the city was thrown.

The perpetual burning and the numerous crawling things were used by Jesus as suggesting the sufferings of the "cast out" souls: Hades, or hell, was like Gehenna, "where the worm dieth not and the fire is not quenched." (Mark 9:48:cf., Luke 16:19-31) (7)

Prior to the first century B.C., however, the Hebrews had no belief in an underworld or Hell, for they believed that when they died, their legacies were carried on by their offspring, hence the historic Jewish emphasis on producing descendants. It was only toward the end of the Old Testament period and into the inter-testamental period (between the writing of the Old and New Testaments), that the concept of Hell developed any degree of sophistication or consequence.

Similarly the concept of Heaven took its wayward turn to spiritualism about the same time. Heaven in the Old Testament was originally translated in the plural as "heavens" and designated a physical, geographical location beyond our solar system. The heavens to the ancient Hebrews were multi-layered. There is some confusion about exactly how many levels their heavens contained, but it is generally agreed that they contained at least three. The third and highest heaven was the abode of the Lord and his angels. The second encompassed the visible stars and planets, and the first heaven was the sky and atmosphere directly above.

This simple scheme corresponds to our universe, stellar neighborhood and sky quite accurately. If the third heaven is considered the universe, then indeed that would be the abode of the Starmen (mistaken for gods). Jesus spoke of Heaven as His Father's house, *"a house with many rooms."* Interpreted from this point of view, such a statement could very well mean that God's universe has many other planets which are inhabited as is Earth. In other words, what is now largely thought of as another dimension or even a spiritual realm was once plainly understood as a *real physical* place. More on this later. Presently, however, it is important to note that Heaven and Hell no longer signify specific geographical locations as much as they denote a state of mind or an intangible existence in after-life.

The shift in emphasis from the present tense and physical understanding of the concepts of Heaven and angels to a future tense, spiritual orientation also has a logical explanation under the Starman Intervention Theory: A mission of extraterrestrials visiting Earth during Old Testament times graphically demonstrated that a physical god and heaven existed. When the mission departed, it left behind a religion replete with the terms "God" and "Heaven," but no physical evidence. It is thus natural to expect that such terms would be perpetuated by the religion in an abstract form thereafter.

Hell and Satan, on the other hand, did not exist as such during that same period but proved to be logical complements to the notion of a "spiritual" God and Heaven. In keeping with the primitive beliefs of other cultures with which the Hebrews had contact over time, the concepts of good and evil spirits found their way into the early church only after their physical counterparts were long gone.

Armageddon, Judgment, Salvation

Armageddon is a term which probably came from the Hebrew "Har Megiddo," or Mountain of Megiddo. This area of Palestine has historically been the scene of violent conflicts and is used in the book of Revelation to symbolically represent the final battle between the forces of good and evil at the end of these supposed trials. At that time all men, both living and dead, will be *judged* according to their relationship with Jesus. Those who accept Christ as their saviour or their redeemer from earthly sin will be translated to Heaven while all the others will be sent to everlasting Hell. The key to gaining entry into Heaven is the acceptance of the notion that Jesus died for the sins of man. This acceptance is the *salvation* and is the focus of Christianity. It is every Christian's duty to actively spread "the word" to the ignorant. The Armageddon, or the end of times of suffering for all good Christians and the beginning of eternal punishment for all unrepented sinners, is the time limit for which mankind must be prepared.

But this time limit has no specific date. It is said to be known only to God and could come at any time. This end, which is supposed to be "at hand," has been preached for the last 2,000 years. Each generation has been told the same thing: that events now taking place in the world signal the imminent collapse of our present system and that we must be prepared to face judgment at any time.

The urgency of this calling swelled the ranks of Christians to the point that it quickly became one of the world's greatest religious movements. In view of its various religio-bureaucratic mechanisms, such as the aforementioned "fear factor," the emphasis on proselytizing, and the unspecified time limit, the Church has the machinery to exist indefinitely. But these concepts too have a relatively late entrance into biblical theology and were not present during most of the Old Testament times. What do secular experts say about the notion of a future Armageddon?

The Revelation of John is the most widely quoted source of information about the "end of times"—although the books of Ezekiel and Daniel certainly fit this form of writing which is called apocalyptic writing—and often speaks in mystical, symbolic, even (to the lay reader) incomprehensible language, of future events. James L. Roberts points out the particular style of this writing:

The characteristics of apocalyptic writing are fairly well known at the present time. In addition to the books of Daniel and Revelation which are part of the Bible there is a wealth of apocalyptic writing in the books of the Apocrypha and the Psuedepigrapha of the Old Testament. A careful study of these writings shows that they have a number of common characteristics. They are produced in times of crises, they describe the conflict between the forces of good and evil, future events are made known through dreams and visions, the end of the conflict is to come shortly, and those who remain faithful through persecutions and trials are promised a reward in the messianic kingdom soon to be established. The messages are for the benefit of the ones being tried and they are usually conveyed by means of symbols which only the faithful would understand. (8)

As for the message of Armageddon so often brandished by Christians, Roberts says:

It is a mistake to suppose that John was predicting events that would take place in the later centuries of Christian history. He was writing to the people of his own day about events that would happen while they were still living. He states that Christ will return while those who put him to death on the cross are still living. (9)

It appears that this fostered sense of urgency about the end to the present order of things, the Heaven/Hell approach to the after-life, and the heavy reliance on proselytizing are driving a movement which had its roots in a series of events and a culture having little or nothing to do with what is being taught. It will be shown that what was once a physical, historical event has been transformed into a spiritual dogma and a system of psychological manipulation. Such misguided religious fervor is, at best, a waste of good people's time and energy. At its worst, it has proven itself a deadly instrument of destruction both of individuals and armies.

Holy and Profane

The terms "holy" and "profane" refer to physical conditions and were used extensively in the Old Testament to denote things which held a critical significance in the relationship between God and man. While the term "holy" today still roughly equates to something that is "of God," and the term "profane" signifies something that is "unclean," it can be demonstrated that these concepts also evolved from a fundamental misunderstanding of physical phenomenon associated with a Starman operation in Old Testament times.

Holiness has been described as an ethical condition which all Christians seek to attain: *Christian holiness is the quality of life and character which comes from being "in Christ"*...(10). But the term as used in the Old Testament clearly represented a *physical quality* of some object or the forces associated with it.

Holy things had great significance to the Hebrews because *they were often dangerous.* The holiness of a thing implied that some force was present which could maim or kill, and God warned of such consequences several times. One example is God's warning to keep clear of the Ark of the Covenant when the priests carried it among the people of Israel: *"...there shall be a space between you and it, a distance of about two thousand cubits; do not come near it."* (Josh. 3:2) The Ark held within it strange powers and was used in conjunction with several wonders, including the halting of the flow of the Jordan River and the toppling of the walls of Jericho. It had such power that the priest Uzzah was instantly killed when he unthinkingly reached out his hand to steady the Ark after it nearly fell while being carried on the march. (II Sam. 6:6-10)

This example suggests that holy things not only had great power, but that the Lord *could not directly control* the power they possessed or conducted except by isolating them from the masses. This idea is further supported by the elaborate measures taken to separate the "holy" things from the masses via a highly trained priesthood. One of God's first tasks in the wilderness of Sinai was to designate a specific tribe of Israel to be "priests," who were directed to wear special clothes, tie back their hair, perform a highly structured ritual in a very specific manner, and handle holy objects in clearly defined ways. Failure to strictly observe these conditions resulted in more than one death. For example Nadab and Abihu, sons of the High Priest Aaron, died when they attempted to offer "unholy fire" before the Lord in the Wilderness. (Num. 3:4)

We shall discover other such accounts as we examine the Bible under the Starman Intervention Theory. As we read the accounts of holy and profane things, consider that the term "holy" signified potentially dangerous objects or forces which were deemed critical aspects of the Starman operation while profane things were things not acceptable to the Lord. Further note that the rules for the use of holy things were carefully defined and monitored by specifically designated "priests" and were strictly off limits to the masses.

Miracles

What is a miraculous event? *Merriam-Webster Dictionary* defines it as *"an extraordinary event manifesting a supernatural work of God."* (11) But the application of the word "miracle" can be a very subjective thing for what is clearly a miracle for one individual at one point in time or space may very well be a common-knowledge occurrence elsewhere. *Harper's Bible Dictionary* defines miracles as *"events in the physical world outside the operation of the known laws."* (12) This is a more reasonable definition because it accommodates the relativity of knowledge and perception in the recognition of a miraculous event.

For the thesis of this book, we must make a distinction between the traditional Christian view that a miracle is a supernatural overriding of natural laws by God and allow for the possibility that biblical works designated as "miracles" can be explained in terms of the difference in technological sophistication between the Starmen and their primitive hosts, the Hebrews. The main issue is, of course, the perception of God either as a spiritual force or as a living being subject to the laws of nature. While the secular arguments over miracles have centered around terrestrial explanations, such as the freak occurrence of a natural phenomenon, the Starman Intervention Theory introduces new factors into the evaluation of miracles. The miracles of the Bible can be divided into three rough categories. The first includes healings, retrievals from death, and the correction of congenital defects in people. The second includes fictional accounts, legends and after-the-fact embellishments, and the third deals with events perceived as miracles due to the differences in technology between the ancient Hebrews and the visiting Starmen.

The healings and cures of illness or defects in people are especially prevalent in the New Testament and revolve around Jesus and his disciples. These involved a wide variety of sensational reversals of maladies ranging from restoring sight to the blind to raising the dead. It is impossible to defend or deny the legitimacy of these works on the basis of the scant biblical evidence alone. But it must be noted that Jesus, though an acknowledged historical figure, was subsequently glorified and deified by all Christian denominations. Any medical skills or healing abilities may have attained new heights through his deification process and may be one possible explanation for his sensational ability to actually raise the dead or cure such severe congenital defects as a withered hand. (Matt. 12:10)

Another possible source of his healing powers which, though known to western cultures, is not as widely accepted as in the East, is known as "faith healing" or "laying on of hands." The study of the powers of the mind involves a vast area of unknown secrets which await man's exploration. We have yet to discover the full extent of the brain's capabilities or where its true limitations lie, but we know that we utilize only a fraction of our available brain capacity at any one time. The power of the mind to control the seemingly involuntary functions of the body has implications for understanding the miraculous healings of Jesus.

One intriguing incident occurred when Jesus returned to Galilee to find a crushing crowd waiting to hear what he had done for others on his travels and some to plead their cases to him. One woman who had been subject to "bleeding for twelve years" sought out Jesus, who was preoccupied with the crowd which, though close, had apparently kept a respectable distance. Upon approaching Jesus, she reached out and touched the edge of his cloak, and "immediately her bleeding stopped." Interestingly, without His seeing the woman, Jesus knew someone had touched him and asked who it was. His disciples blamed it on the pressing masses, but Jesus persisted in knowing, saying a curious thing: He said, *"Someone touched me; I know that power has gone out of me."* (Luke 8:46) Jesus had some power which he could transmit to others, either through conscious act or by the faith of believers in him, that perhaps produced healing.

Some relatively primitive but fascinating experiments are being conducted by groups of serious students in an emerging science called parapsychology. Enough solid evidence has emerged from these experiments to show that the body can be consciously controlled by the mind. One such technique called biofeedback is already becoming a widely used means of reducing high blood pressure, migraine headaches and tension. Another discovery has revealed that the body, and in fact all living things, contain and emit an energy field which can be photographed and interpreted. The process is called *Kirlian Photography* after the Soviet scientist Semyon Kirlian, who discovered it in 1939. Recent research widely published indicates that these photographs of the body's electrical field can be used to detect a person's mood and even the relative state of health! Some proponents believe these energy fields are the forces which the brain directs to produce healing in the body, among other things.

The Soviet Union has long since abandoned its ridicule of such ideas and has spent a great deal of energy officially investigating these and other psychic phenomena for both scientific and military applications. The complexity and scope of this subject prohibit lengthy discussions in this book. However its significance and implications regarding the miracles of Jesus must not be overlooked.

Much of the thesis of this book is based upon a logical projection of our current scientific knowledge *back* into the past to analyze the events of the Bible in the context of a technologically advanced contingent of extra-terrestrial beings operating in a non-technological earth environment. If we have the ability to control the body by consciously directing the proper internal forces (however crudely), there is a reasonable basis for expecting that we may someday practice such techniques routinely on much more sophisticated levels.

Similarly, if Jesus was the "Son of God," viz., a Starman, such techniques could also have been employed by His advanced culture in the relatively primitive Roman environment. The topic of Jesus will be covered more extensively in a later chapter, but for now it is being suggested that a strong faith in the abilities of Jesus, coupled with the mind's ability to direct the body's energy to heal itself, is a possible explanation for some of the miraculous healings of the Bible.

A second classification into which biblical miracles might be grouped is that of fictional legends, which are rooted in the magic, rituals and lore of other peoples and other times. This is clearly the case, for example, in the account of the ten plagues which eventually forced Egypt's pharaoh to set Israel free before the Exodus. The final version of the plague story has *three* sources which originate in different times and places in Israel's history. Of the three sources (called the J Document, the E Document and the P Document), J is the oldest and records only seven of the ten plagues. According to *Harper's Bible Dictionary* (13), all of the plagues described are consistent with the Egyptian setting. Hebrew tradition is seen to have recorded ten severe events of nature which were particularly punishing to the Egyptians and later attributed them to the hand of God as the reason for Israel's freedom. Other miracles in this category would include Moses' rod turning into a serpent (Ex. 4:2-4) and perhaps the account of the fiery oven into which three men of God were thrown but were not harmed. (Dan. 3:12-20)

The third classification concerns events which were certainly miracles to the Hebrews in the day in which they occurred but which become quite clear, in the light of our present knowledge, as a technologically advanced contingent of alien beings working amongst a less sophisticated indigenous one. Much attention will be given to redefining the nature and purpose of such events as the parting of the Red Sea, the halting of the flow of the Jordan River, ceremonies associated with the

Ark of the Covenant and service to God, the wars of God, the fall of Jericho, and other key occurrences and how they apply to the theory of a Starman intervention into the Hebrew history.

One might ask at this point why extra-terrestrials would choose to come to this planet. Perhaps that can best be answered by simply looking at our own development. Why do *we* reach for the heavens? We do so to discover, to better understand our universe, to meet the challenge, and for each of these reasons and more! Eventually we will expand the radius of our reach to include other stars and galaxies, for there will be new worlds to explore and new mysteries to discover.

When we come upon a world inhabited by less advanced beings, what will our ethical standard be? Will we intervene intellectually or biologically out of compassion for their pitifully "backward" state? Will we exercise the Manifest Destiny mentality of the early missionaries, who saw it as their duty to bring civilization and God to the savages? As God said to his people in the book of Psalms, *"I will instruct you and teach you the way you should go. I will counsel you with my eye upon you."* (Ps. 32:8) In so doing we may establish an outpost colony, train the savages and even take wives as we see fit. And we would likely be duplicating an ancient process which spans the universe. Jean Sendy proposed that *"...we are only a link in a chain of civilization...an organization extending throughout the galaxy."* (14)

And so perhaps the natural order perpetuates itself on a grander scale than yet imagined. But is this not more logical, rational and conservative than some legends of distant miracles worked by ethereal winged angels? As it appears now, the story of the Bible does not quite fit in with what we know and experience in life. There is the allusion to a *significantly different* order whose parts don't quite mesh with what we intuitively know about the laws of nature and how a supernatural God would deal with flesh and blood subjects. For example, why would God choose to part the Red Sea so a chosen people could cross over, away from danger? Why not simply change the *minds* of the Egyptian pursuers or even just "blink" them out of existence? In other words, the whole nature of the Old Testament stories is on a very physical level. The sum of the evidence points not to a supernatural override of natural laws, but a sophisticated physical presence in a relatively primitive environment.

A race of hybrids created by the inter-marriage of Starmen and Earthmen would be of special importance to the Starmen for they would be the *literal* sons of God. Such offspring would be afforded special attention over all the other races, that they might be established on the earth. As the Bible states, the seeded culture would then become "priests" to teach the rest of mankind to become an ethical society. In the Starmen's own best interest, the people could then eventually move out into space with some semblance of moral authority to join the interstellar confederation of many worlds. This is the story which is recorded in the Bible.

Endnotes

5. Madeleine L. and J. Lane Miller, *Harper's Bible Dictionary*, Harper and Row Pub., New York, 8th Ed., 1973, p.647.

6. According to *Harper's Bible Dictionary*, the name "Lucifer," meaning "light-bringer," "was a name given to the planet Venus when it is the morning star. It is used in Isa. 14:12 A.V. to render the Heb. 'shining one' applied to the King of Babylon, fallen from his high estate. In the 3d century A.D. the saying of Jesus: 'I beheld Satan as lightning fall from heaven' (Luke 10:18) was erroneously supposed to refer to Isaiah 14:12. Hence Lucifer came to be regarded as the name of Satan before his fall." *Ibid.*, p.402.

7. *Ibid.*, p.240.

8. James L. Roberts, Ph.D., *Cliff's Notes on the New Testament*, C.K. Hillegass, Pub., Lincoln, NE, 1965, p.92.

9. *Ibid.*, p.93.

10. Miller, *op. cit.*, p.265.

11. *The Merriam-Webster Dictionary*, Pocket Books, by G.&C. Merriam Co., New York, 1974, 12th Printing, p.447.

12. Miller, *op. cit.*, p.447.

13. *Ibid.*, pp.562-3.

14. Jean Sendy, *Those Gods Who Made Heaven & Earth*, Berkley Medallion Books, New York, 1972, p.75.

God Creating the Universe by William Blake

Chapter IV
EVOLUTION AND THE BIBLE

Sceptical scrutiny is the means, in both science and religion, by which deep thoughts can be winnowed from deep nonsense.
- Carl Sagan

There are two very important reasons for looking at the subject of evolution and the Bible before proceeding. The first is that it is critical to the thesis of this book to show that the events related in the Bible are reconcilable with both Christianity and science to demonstrate that the Bible tells a story that is by and large historically true, even though its message has been distorted over time through man's misunderstanding.

The second reason is that if it can be shown that the process of evolution is a universal phenomenon, then it will become easier to believe that such a process *also gave rise* to the alien culture which is claimed to have visited us so long ago. By demonstrating that the same laws which gave rise to us also gave rise to the Starmen, important common traits and characteristics will become more readily apparent and the logic of the Starman Intervention Theory much less tenuous.

Key Arguments

The biblical creation epic has always been a point of contention between evolutionists and creationists, primarily due to the belief that each side held the correct approach to understanding how man came into being. If it can be shown that evolution and creationism are compatible, then at least one major barrier to understanding the Bible in its proper context can be overcome.

When we first open the Bible, we read in Genesis of the creation of the universe. Many cultures of the world have creation epics, to be sure, but a rather startling thing about the biblical account is that it is becoming recognized as being quite consistent with current scientific views on how the universe came to be. *Time* magazine's Lance Morrow perhaps said it best:

The so-called Big Bang Theory makes some astronomers acutely uncomfortable, even while it ignites in many religious minds a small thrill of confirmation. Reason: The Big Bang Theory sounds very much like the story the Old Testament has been telling all along (15).

But the crude account found in Genesis has been assigned dates as early as 1450-1410 B.C. and earlier. How can we account for this? Recall that we earlier proposed that there were actually two separate forms of God represented in the Bible: the "creator God" as manifested in the natural order of things, and the distinctly human-like God who walked among men. It is in Genesis, chapter one, that we encounter the creative force akin to the natural laws which contemporary science seems to be confirming. In fact it was probably the second form of "God," (the Starmen) who gave the Hebrews this story of the creation by the *real* God. But was this creation a single act, as claimed by Christians, or was it a long process, as science claims?

Some of the major problems seem to have arisen over the issue of semantics. If God created the universe, could He not have done so through the evolutionary process? When we read that the universe was created in six days, is it possible that "days" was used for illustrative purposes? Time to God is of little consequence. In II Peter 3:8 we read, *"But do not ignore this one fact, beloved, that with the Lord one day is as a thousand years, and a thousand years as one day."* In this sense, we might think of the term "days" as referring to periods of time.

Similarly we read in Gen. 2:7 that man was created from the dust of the earth. This is more correct than at first might be apparent, for man's evolutionary origins are indeed of the "dust" of the earth: compounds and chemicals needed to form the first life were very elementary and simple. These portrayals of evolution are universal in meaning. They have served to illustrate the basic evolutionary sequence for the ancients as well as to allow modern man to discern their basic correctness. The ancients would have had little or no use for a passage that reads, "The earth was created some 4.5 billion years ago." A *billion* is a mind-boggling number which we let computers handle today. Likewise any discussion of the mechanics of the genetic process would have been grossly out of context in the fifteenth century B.C.

Thus the disputes between science and religion have yielded an interesting situation. What has been locked in religious verse for thousands of years is just now being confirmed by science. The once special position which creationists felt they occupied in the universe has been shown by science to be quite insignificant, indeed, among the countless stars of the universe. And yet scientists for centuries smirked at the creationists' views of how our universe came to be until it was suddenly realized that the order of the Bible had just been scientifically verified!

Several of the Christian arguments against secular theories of evolution center around the mechanics of the evolutionary process. For example, there is a common misconception among Christians about the effects of evolution on living and non-living things. What has happened is that the creationists have confused two different forces in nature: biological evolution, which is governed by behavior and environment under a rule called "survival of the fittest," and physical evolution, which is governed by the Second Law of Thermodynamics, often called entropy.

Basically, after the universe was created, it began to change under a set of physical laws which cause it to move from an unstable, complex state to increasingly more simple, stable conditions in a gradual process of "wearing down" over immense periods of time. Living things, on the other hand, are not governed directly by these same laws because they grow and adapt to changes in the non-living world around them. As the climate or the physical conditions of their environment change, they must modify their behavior, structure or location. Failure to do so inevitably results in the extinction of the species. Indeed the massive fossil evidence of extinct species is ample testament to this law. This confusion by creationists is reflected in a quote from a religious book contesting evolution:

Is there any trend toward evolution among the elements on earth? ... Will the elements of the earth, left to themselves ever produce an automobile, or even a simple gear? ...Inanimate matter on earth simply does not search out a way to improve itself, but tends toward a state of neutralization or stability. Nor will appeals to immense periods of time help. (16)

In one sense this statement is correct: the elements of the earth, in the state that they are in now, do not consciously attempt to build things. Nor will we probably ever see new life form in this day and age, for the special conditions required for the creation of life as we know it have long since disappeared because of the very changes in the earth that were quoted above.

The earth today is nothing like the primitive environment before life on earth. In fact life as we know it today could not have existed in that violent pre-life world. It was characterized by great quantities of deadly radioactive materials, ammonia, methane, carbon monoxide gas and extreme temperatures. Because the earth does change, life today is quite different from life even one million years ago. By understanding these basic tenets of evolution, the pictures painted both by science and Christianity can be reconciled.

Science still cannot explain what happened before the "beginning." The question of the "prime mover" or the "it that became" is still left to philosophers and theologians, and that is where it may stay for some time. But anything beyond the physical, measurable universe is irrelevant anyway since there can be no verification, and the theory of this book can be covered by what is already known and verifiable. Therefore, when the universe was born, time began, and that is where we must start.

Geography of Space

To better understand our insignificant position in the universe, we must first have a basic knowledge of its geography. On a clear night, we can look into the heavens, just as the ancients did, and see perhaps 4,000-5,000 stars with the unaided eye. But with the aid of giant telescopes in modern observatories, literally *thousands of billions* of stars come into view. Each star is a sun, like our own, which may or may not have planets revolving around it. Our sun is a middle-aged star of rather ordinary size and mass. Our earth is the third of nine planets revolving around the sun. Together this family of planets and sun comprise a solar system.

Most stars are formed into groups called galaxies. Our galaxy is called the Milky Way, and it is so large that scientists must measure its diameter in "light years." A light year is the distance which light can travel in one year. At the speed of 186,000 miles per second, a beam of light will travel 5,880,000,000,000 miles in one year! (17) The Milky Way Galaxy, which may contain as many as 100 *billion* stars, is about 100,000 light years in diameter and about 10,000 light years thick! Even the most powerful telescopes cannot see all the way across our own galaxy, which looks like a giant spiral or pinwheel spinning majestically in space.

Our sun is situated on the outer rim on one of the arms of the galaxy, and we are humbly insignificant in this incredibly expansive universe. There are many of thousands of galaxies beyond ours, and they vary greatly in size and shape. Some appear to be forming while others may be dying in fiery stellar explosions called "quasars," which destroy countless stars in a single terrible sweep. The universe is full of violent beauty and wonder. Fortunately we live in a relatively moderate stellar neighborhood. But the universe is ever-changing. It was born in a colossal burst of energy and it has continued to evolve ever since.

We know the universe is evolving because atronomers have measured the distance between various stars and have discovered that nearly all of the galaxies in the universe are moving *away* from each other. This was determined by analyzing the light emitted by each star in a process called "spectral analysis."

When the light from an object is passed through a special prism, it is refracted into its component wavelengths, which we see as a band of colors. The shortest visible wavelengths are violet. After violet come blue, green, yellow, orange and red, the longest visible rays. The light from a stationary object, when passed through a prism, is always refracted into the same pattern. If, however, the object is moving toward the observer, the pattern of colors will shift toward the blue and violet ends of the spectrum. Conversely, if the object is moving away from the observer, this pattern will shift toward the red end, with the longer wavelengths. These patterns are fixed according to the object's speed and thus are a valuable tool in determining stellar velocities.

It is this red shift in the refracted color patterns which verified that the galaxies are all moving away from each other at a relatively fixed rate of speed. Tracing the paths of the galaxies backward, all intersect at a common point. Knowing their velocities, it has also been determined that all galaxies came from this common point about 15 billion years ago. (18)

If all galaxies originated from a common point, then the universe must be composed of the same materials. To check this, scientists again turned to spectral analysis to study the properties of the light emitted by stars. In 1814 a German optician named Joseph Von Fraunhofer (1787-1826) found that when he passed beams of light through some of his fine glass prisms, they produced the characteristic spectrum, but there was a difference.

Instead of a continuous band of colors, each color was separated by dark lines of various widths. Today these lines are called "spectral lines" or sometimes "Fraunhofer lines." They are distinctive in pattern for each element, and it is in this manner that astronomers can determine the physical properties of distant stars. (19)

Using spectral analysis, the sun's composition was first crudely deduced by the Swedish astronomer Andres Jonas Angstrom in 1862. Later, when the process of spectral analysis became more refined, it was discovered that not only composition, but quantities could be determined. According to Asimov:

> *The recent estimates of the American astronomer Donald Howard Menzel (1901-) show that 81.76 percent of the sun's volume is hydrogen and 18.17 percent is helium. That leaves only 0.07 percent for all other atoms.* (20)

This is corroborated through the analysis of meteorites and our moon samples, which show that all of the universe as we know it is composed of the same basic materials in varying quantities. It was a Belgian astronomer named Georges Edward Lemaitre who first proposed the idea of a gigantic "cosmic egg," as he called it, which exploded and sent the building blocks of the universe hurtling into space. Other astronomers saw merit in the idea, examined it and eventually espoused it as the model of the birth of the universe, which is today called the Big Bang Theory.

In sum this evidence, and much more which goes beyond the scope of this book, suggests that there was indeed one big explosion which sent huge quantities of stellar material shooting out in all directions in the beginning of what we know as the universe.

All things taken together, the idea that the entire universe is composed of the same materials, that it operates under the same laws of nature everywhere, and that we are an integral part of that universe, points out two important things: First, that we are products of the evolution which is occurring all around us and, second, that because evolution is a universal phenomenon, we are probably not the only form of life in this magnificent order of things. From this secular perspective, then, let us examine what the Bible says about creation.

Redefining the Creation

"In the beginning God created the heavens and the earth." So begins the Bible. What existed before the beginning, we cannot know. What can be disputed is how God works. The very abbreviated version of the creation found in Genesis can be summarized as follows:

1. *The earth was without form and void, and darkness was on the face of the deep*

2. *And the spirit of God was moving on the face of the waters*

3. *And God said, "Let there be light"...and God separated the light from the darkness*

4. *And God made the firmament and separated the waters which are under the firmament from the waters which are above the firmament*

5. *And God said, "Let the waters under the heavens be gathered into one place"*

6. *"and let the dry land appear"*

7. *and God said, "Let the earth put forth vegetation"*

8. *And God said, "Let there be lights in the firmament of the heavens to separate day from night;"*

9. *And God said, "Let the waters bring forth swarms of living creatures"*

10. *"...and let the birds fly above the earth"*

11. *And God said, "let the earth bring forth living creatures according to their kinds; cattle and creeping things and beasts of the earth according to their kinds"*

12. *Then God said, "Let us make man in our image, after our likeness"*

These passages, taken from Genesis 1:1-26, are from the Revised Standard Version of the Bible. The format we shall use for analyzing the compatability between Christianity and scientific views of the creation will follow a basic model laid down by Peter W. Stoner in his delightful little book *Science Speaks* (21), although it must be noted that Mr. Stoner's work was oriented toward a different audience.

To understand how close the Genesis passages come to the scientific understanding of the formation of the universe, we begin by looking at the supposed primordial blast - the beginning as we know it. Early theories about the formation of the universe caused deep disagreement between the Genesis account and the established views of astronomers.

A theory called the "Nebular Hypothesis" was advanced by French astronomer Pierre Simon Laplace (1749-1827) during the nineteenth century. It held that the solar system started out as a whirling gaseous mass of extremely hot material. As this mass cooled, it shrank and flattened into a disk shape. Turbulence at the rim of the disk was supposed to have produced smaller whirling masses which formed the planets as they cooled, or so it was thought. But the idea proved defective when physicists showed that hot gases would not react in that way at all.

Several new theories were advanced, the most prominent being the "Planetesimal" and the "Tidal" theories. The Planetesimal theory envisioned two stars coming into close proximity to one another so that their respective gravitational fields would pull swirling masses of star material away from each star and out into space, thereby cooling the masses down enough to form planets. Neither theory, however (the Tidal theory being similar in several respects), could adequately explain how the super hot gaseous material would coalesce into planets:

> *In 1939, the American astronomer Lyman Spitzer, Jr. (1914-) advanced what seemed valid arguments for concluding that such super-hot matter would never condense to form planets but would spread out rapidly to form a gaseous nebula about the Sun, and would stay so.* (22)

By studying the galaxies, astronomers were also coming to the conclusion that collisions and near misses among stars are rare occurrences, and there was good evidence that many stars, indeed perhaps most stars, have planets revolving around them. So astronomers abandoned the Planetesimal and Tidal theories and returned to an old idea which was predicated on a contracting nebula. What they discovered upon closer scrutiny of the available evidence was the "Diffuse Nebula," a phenomenon which corresponds strikingly to the first verse of Genesis:

The earth was without form and void, and darkness was upon the face of the deep.

A Diffuse Nebula is a great cloud of rarified gases, mostly the simplest gases such as hydrogen and helium, which are collected together in space through the effects of their own gravities and by the various forms of radiant energy given off by other stars, such as light, x-rays and gamma rays. There are many thousands of these formless nebulae visible through telescopes, and it is thought that they are nearly as common as stars.

At first astronomers thought they were simply areas of a galaxy with no stars. Some even proposed that they might be holes in space. But their plumes have frequently been observed to be intruding into bright clusters of stars so that they become silhouetted, and it is precisely this phenomenon which prompted German astronomer Max Franz Joseph Cornelius Wolf (1836-1932) to propose that these masses were actually dark clouds of material and not just holes.

Two of the most famous of these are the Horsehead Nebula in the Orion system and the North American Nebula in Cygnus. It is in this billowy stellar material that we find the building blocks of the stars and planets and the dark beginning "without form and void."

And the spirit of God moved upon the face of the waters

"Spirit," in the sense in which it is used in this verse, comes from the original Greek word "pneuma" or, literally, "wind." We know that as these dark clouds draw into themselves and condense, there is turbulence and heat. Large nebulae (Latin for clouds) moving about in a state of slight vacuum tend to produce a red glow as gravitational pressures build up at the center of the cloud. As this condensing continues, the gases near the center of the cloud become a superhot plasma that actually begins to burn.

In order for such a cloud to give birth to a star, it must contain enough material to give it a mass from one-fifth to five times that of our sun. Otherwise it will not exert enough gravitational pressure upon itself to produce the heat necessary to cause fusion reactions which would turn it into a blazing star.

In the turbulence of this pre-star stage, the plasma must maintain its integrity. If it loses some of its mass or if some smaller masses in the vicinity have sufficient momentum to stay clear of the gravitational force of the main body, they will cool to become planets.

And God said, "Let there be light"...and God separated the light from the darkness.

In their primordial states, some one billion years before life, the planets of our solar system began to cool off and form into solid masses. The sun continued to shrink because of the fantastic gravitational forces of its mass which keep it burning. Eventually the planets became fixed in their orbits. The particular distance at which each planet became fixed determined its conditions regarding whether life could form or not. In our solar system, it was the Earth which became fixed at precisely the right distance from the sun to allow the necessary conditions for the development of our pre-life environment.

As the Earth revolves around the sun, it also rotates about its axis so that there is daylight on the side of the earth that faces the sun and darkness on the side that faces away. The same spot on the Earth will pass the sun approximately every 24 hours, giving the earth a moderate surface temperature. Some planets have faster rotations, such as Jupiter with a rotation of 9 hours and 50 minutes, and some have slower rotations, such as Venus, which rotates counter to the Earth once every 243 days! (23)

And God made the firmament and separated the waters which are under the firmament from the waters which are above the firmament.

In the beginning, hot gases, toxic fumes and clouds of ammonia and water vapor surrounded the earth. Volcanic activity spewed molten lava about and sent tons of ashes skyward. Intense sunlight, cosmic rays and streams of electrons from the decay of radioactive materials bombarded the Earth's surface forming the early compounds and water vapor. As the Earth cooled, water vapor in the sky condensed on the volcanic ash floating in the air, causing rainfall and violent electrical storms. More advanced gases formed, such as nitrogen, hydrogen sulfide, methane, ammonia, carbon monoxide and carbon dioxide. Oxygen was also forming, but it combined with other substances so readily that there was never much free oxygen left floating about. Because the earth was still quite warm, a large portion of the water remained suspended in steamy clouds, which formed a thick blanket around the planet.

And God said, "Let the waters under the heavens be gathered together into one place."

As the earth cooled, more and more of the water vapor condensed into rain and fell to the ground. Rain began to fall constantly. Great rivers formed and flowed into great pools, which became lakes and seas.

"And let the dry land appear."

Scientists know that as the earth continued to cool, it tended to shrink somewhat, which buckled and bulged its outer crust causing violent earthquakes and volcanic activity. We still experience earthquakes and volcanic activity today in various parts of the world as great sections of land called "continental plates" continue to shift. Entire mountain ranges pushed skyward, and with them arose entire continents. Thus the once smooth earth became continents and oceans.

And God said, "Let the earth bring forth vegetation."

Plants dominated the earth long before animals, but the first types of life are difficult to classify since they are thought to have had traits of both plants *and* animals. Species of diatoms still exist which have those same characteristics.

In laboratory experiments simulating the primitive atmosphere of the pre-life Earth, scientists have succeeded in creating many of the compounds necessary for the formation of the first simple life forms:

> *The formation of monomers from the gases of the primitive atmosphere is the step about which most is known, since the reactions can be simulated and studied in the laboratory.... It is clear that amino acids would have been readily synthesized in the primitive atmosphere.* (24)

In similar experiments containing these nutrient chemicals, scientists have also observed the formation of enclosures of cell-like membranes. Several functions of living organisms have been mimicked as well, including a growing process, photosynthesis, simple metabolic functions and even reproductive sequences. It is now quite clear how the first delicate life arose from that ancient primordial soup some three billion years ago.

The point that separated actual life from blobs of cellular material was a metabolism or the process of absorbing food to sustain animation. The earliest metabolic process appears to have been anaerobic (not requiring oxygen) fermentation, in which single-celled organisms break down and absorb organic compounds after surrounding and engulfing them.

The first cells, which probably resembled a modern bacterium called *Clostridium,* fed exclusively on non-biologically formed energy-rich molecules to survive. They only consumed energy and generated none of their own. For a time life was restricted to this type of life form because of the limited supply of such energy-rich molecules.

But with the help of these first organisms, the carbon dioxide content of the atmosphere was raised from one to three percent. Carbon dioxide absorbs the sun's energy and converts it to heat. The warmer the atmosphere became, the more clouds formed, which effectively reduced the amount of sunlight that reached the surface of the Earth. As a result, fewer inorganic compounds could form, and a new way to survive evolved, called photosynthesis.

The new organisms could use sunlight and the chemical chlorophyll to convert simple proteins into energy. This greatly enhanced their chances of survival by removing them from competition with other organisms for the dwindling supplies of inorganic compounds.

They also began to use carbon dioxide and gave off a new waste, oxygen. Eventually enough oxygen was added to the atmosphere so that an ozone layer could form around the earth, thus blocking out the ultraviolet radiation which sythesized the inorganic compounds used by the scavenger bacteria. Plant bacteria thus came to dominate the Earth, and in so doing completely changed the environment. Never again would new life spontaneously generate on the Earth.

And God said, "Let there be lights in the firmament of the heavens to separate day from night."

Photosynthesis completely reversed the process of adding carbon dioxide to the atmosphere. This had the effect of lowering the temperature somewhat, allowing more moisture to condense into clouds and fall to the Earth as rain. Breaks in the clouds allowed the sun to shine through during the day and the moon and stars to shine through at night.

A hardier type of vegetation soon developed which could withstand the direct rays of the sun and the greater variation in temperature which accompanied the breaks in cloud cover. The rise in the level of oxygen in the atmosphere also stimulated the development of organisms which could tolerate it. Eventually it became vital to sustaining life.

About this same time in the development of unicellular organisms came a new way of reproducing, called sexual reproduction, which produced a greater genetic variation from parents to offspring. According to J. William Schopf, *"Because sexual reproduction allows novel adaptations to spread quickly through a population its development accelerated the pace of evolutionary change."* (25)

And God said, "Let the waters bring forth swarms of living creatures."

The first single-celled organisms, called *prokaryotes,* reproduced asexually. But somewhere around 1.5 billion years ago, a much more complex cell emerged with greatly enhanced adaptability. What had taken nearly four billion years to develop from the simplest elements had now culminated in the *eukaryotic cell,* the basic building block of all modern living things.

Most of the developments up to this point involved chemical and metabolic improvements in the cell itself, but during the last 500 million years the cell has diversified into every multi-celled organism now roaming the earth, and many, incidentally, which have failed to survive. The development of the eukaryotic cell virtually exploded the evolutionary pace. The advantage of multicellular organisms living in close proximity was amply demonstrated in their survivability, specialization and reproductive efficiency.

It has been said that the sea is the "mother of the earth," for within her watery womb developed the prototypes for all living creatures. From the warm-water shallows where primitive cellular colonies began to congregate came the sea anemones and burrowing tube worms. The fossil remains of their burrows reveal that they were already established by the late Precambrian period, some 700 million years ago. As stationary life forms, they relied on the passing currents to bring their food. From the anemones and tube worms soon branched other diverse forms which could move along the bottoms to search for food and shelter, and from them came still more forms which could move about freely in the water.

The first animals with skeletons arrived on the scene during the Cambrian period, perhaps 550 million years ago, and the first true fish shortly thereafter. Given the vast expanses of oceans and the adaptability of living things, it is not surprising to find that the fishes burst forth in a rich diversity of forms over the next 60 million years. With their superior mobility, they quickly spread throughout the oceans, establishing their individual ecological niches.

During the Silurian, about 400 million B.C., there evolved two distinct branches of fish: the ray-finned and the lobe-finned fishes. It was from the line of lobe-finned fishes that the amphibians came, animals which can leave the water for short periods of time to look for food or move from a landlocked tide pool to the ocean. Prior to the actual emergence of the amphibians, certain of the lobe-finned fishes had acquired the ability to gulp air directly from the atmosphere and move about the shallows and tide pools using their stubby pectoral fins or modified gill plates. Certain of these transitionary species still exist today, like the Periophthalmas (mud skipper) which moves about freely on land breathing air through the moisture trapped in its gill chambers.

The conquest of land was a slow process because it involved not only having strong enough legs to lift the body, but a skin to prevent dehydration, an atmospheric breathing apparatus and a means of producing and rearing offspring in a dry environment as well. While the amphibians produced some formidable advances in land mobility and breathing, they never completely broke with the need for a watery environment, chiefly for reproductive reasons.

Then, about 300 million years ago, there developed a way to provide the watery environment for the early stages of an offspring's development—the amniotic egg. It had a hard outer shell which could be laid on land, and it marks the birth of the age of the reptiles, the first true cordates to break their ties to the water.

The reptiles quickly radiated out, dominating the earth in every conceivable size and shape for nearly 165 million years. From these reptiles evolved the dinosaurs, including the Tyrannosaurus, a voracious meat-eater which stood 20 feet tall and was over 50 feet long from head to tail! But they also produced the small, swift and intelligent forerunners of the birds and mammals. They diversified so completely, in fact, that some members even returned to the oceans. It is possible that this is addressed in Genesis 1:21, which reads:

God created the great sea monsters and every living creature that moves, with which the waters swarm...

Good examples of these "sea monsters" are the Icthyosaurs and the Plesiosaurs, which lived 200 to 100 million years ago. According to Jarman, some were quite formidable:

The most savage reptiles were the mosasaurs and geosaurs. The geosaurs were marine crocodiles of the Jurassic and lower Cretaceous. The mosasaurs seem to be confined to the Cretaceous. Some grew to 30 feet long. Many of these aquatic reptiles had long snouts and sharp teeth, which were used for eating fish. (26)

Recently there has been some speculation as to whether or not the great dinosaurs were warm-blooded. New views on this are supported by the posture and carriage of specific members which suggest they were very agile, especially the carnivorous varieties. Not only were they probably more agile and active than once thought, they may have been smarter and more alert too.

"...and let the birds fly above the earth...and every winged bird according to its kind."

The ancient ancestors to the birds are small lizard-like reptiles that walked on their hind legs and laid eggs. One such ancestor may have been Psuedosuchain, which lived nearly 270 million years ago. From this pre-bird stock branched at least two completely different forms of winged animals about 135 million years ago: the Pterosaurs and a lineage of true birds. The Pterosaurs took various forms, including the Pterodactyls, which had wingspans of up to 25 feet. These were gliders and soarers as opposed to active fliers, however, and merely paralleled the development of another line which had feathered wings and looked much more like the birds we know today. The first true bird of which we have fossil evidence is the Archaeopteryx from the Upper Jurassic era, about 100 million years ago. It was first discovered in Bavaria in 1861.

Although the skeleton resembles that of a reptile more than that of a modern bird, the archaeopteryx had feathers and also wings for flight. Because its claws were well developed, the archaeopteryx is believed to have spent most of its time climbing trees, occasionally gliding from branch to branch or from a tree to the ground. (27)

Evolutionary changes are often brought on because of changes in the behavior of the organism. Evolutionists have speculated that the development of flight came about when tree-dwelling pre-birds jumped from trees. Their jumps favored the development of a mechanism which could slow the animal's descent much the way the parachute-like body structure of the flying squirrel (Rodentia sciuridae) helps it to glide as much as 200 feet from tree to tree. (28) Another theory advances the idea that running reptiles developed wings to aid in attaining speed across the ground to escape enemies or to catch prey. Eventually they began to lift into the air for progressively longer flights. (29)

And God said, "Let the earth bring forth living creatures according to their kinds; cattle and creeping things and beasts of the earth according to their kinds."

The turning point for the terrestrial reptiles was during the latter part of the Cretaceous era, when mammals began to replace them as the dominant vertibrates. Actually there were lineages of mammals during the reign of the dinosaurs, but they remained small in size and played a relatively insignificant role during that time. It has been suggested that the reason the mammals remained in the background was that the dinosaurs themselves were warm-blooded. Therefore while the dinosaurs predominated in the larger varieties, the mammals found niches for themselves in the smaller ones.

But by the end of the Cretaceous, the weather began to change dramatically. It has been suggested that a large asteroid or comet struck the earth, setting off violent earthquakes and global volcanic activity which so altered the environment that its capacity to support large animals was destroyed. The great dinosaurs quickly vanished from the scene, leaving the smaller animals to radiate out and fill the ecological niches left by their larger predecessors.

From among those smaller animals emerged the mammals, which themselves split into branches, including those with incubation pouches, called marsupials, and the placental mammals. With these types of innovations in the reproductive systems, the offspring remained in or near the bodies of the parents, where they were afforded the greatest protection from a hostile environment. In the case of marsupials, the young are born at a premature stage, and are then transferred to a special pouch on the body of the parent for incubation. Placental mammals carry the fetus to maturity. The offspring are born completely developed and ready to follow the parent. The placental mammals seem to have evolved from an early marsupial cousin, *Deltatheridium,* which was a small insectivore that lived in Mongolia during the Upper Cretaceous. He was a weasel-like animal that had characteristics of both insect-eating marsupials and placental mammals.

So successful were the mammals that they rapidly made the far corners of the earth their homes. From the tiny mouse to the giant whale, mammals testify to the survival advantages of placental reproduction and parental training. Indeed man, the master of the world, is a mammal!

Human Evolution

Then God said, "Let us make man in our image, after our likeness."

Here we come to a critical point in the analysis of the evolutionary theory for two reasons: First, it is clear from the Genesis passage above that God is talking of making man to others, creating man in *their* image. The obvious question is why is God referred to as speaking to others? The second reason involves the Christian view that man is exempt from the evolutionary process even if there is a case for the evolution of all other living things because they hold mankind was specially created by God.

First, let us deal with the evolutionary problem. The Bible tells us that God specially created man to rule over all the other living things on the earth and is thought by many to be superior to them. But while his intellectual capacity is certainly higher, biologically man is virtually no different than any other animal, regardless of intelligence. Furthermore, his origins from the lower primates is obvious.

But while our evolutionary roots are to be found in the primate line, to say that we evolved from the apes is not correct. Michael H. Day clarifies this common Christian misconception:

Perhaps the commonest fallacy that exists in the study of evolution is that man arose from the apes. This misbelief has been used many times by opponents of the evolutionary theory. Clearly no living ape can be man's ancestor but, if man's own evolutionary history is being considered, then the history of other living primates must also be examined. Modern apes, however, in no way represent "missing links" in man's evolution. (30)

Let us first explore how man could have evolved naturally, and then we will address the passage in Genesis which states that God created man, and show how it can be reconciled nicely with our existing knowledge of evolution.

The earliest known fossil primates emerged somewhere in the middle of the Paleocene period, about 50 million years ago. In parts of North America and Europe a small squirrel-like insectivore called *Plesiadapidae* emerged which had forward-facing eyes and had probably walked on all fours. Within 10 million years it had split into five or six different variants, including predecessors to the old world monkeys, the lemurs and the apes.

The last few years of anthropological field work have produced so many new and important fossil finds that it will take years to sort and evaluate them all. But enough evidence now exists of man's predecessors, according to Richard E. Leakey, that they can now be traced to as far back as five or six million years with some degree of certainty.

The generally accepted beginnings of true man-like creatures were the *Ramapithecines,* which lived at least 12 million years ago and were distributed throughout Europe, Asia and Africa. From them branched two distinctly different forms, which appear to have been contemporaries: The line of Australopithecines, which itself split into two more lines (one of which was slightly built, the other more robust), and the line called Homo, from which we are descended. (31)

The Homo line seems thus far to have occurred only in Africa. It is clear that he had the more advanced social structure, which made him more durable than his Australopithecine cousins who eventually died out. There is also evidence suggesting that groups of these different variants crossed paths in some regions, so that evidence of two different cultures have been found to exist side-by-side in some parts of Africa. Homo is thought to have been driven into the southern African territories of the Australopithecines by various ice ages. The gradual decline of the Australopithecines could have occurred either through the assimilation of one culture by the other, or more likely, the Homo line simply won out over his southern cousins in the keen competition for limited resources. M.H. Day explains:

The evolution of man from his primate ancestry has taken place rapidly during the last two million years. One of the reasons for this is that during the Pleistocene period the climate was very unstable. Arctic conditions spread down from the poles at least four times bringing snow and ice deep into Europe and North America. In Africa there were corresponding wet phases of high rainfall. Thus, evolutionary pressure in the early hominids was high, and the environment ruthlessly eliminated those forms which could not adapt to the changing conditions. There was therefore a high premium on new and advantageous structural and behavior characteristics, such as mental ability and the use of the hands for making tools and weapons. (32)

We owe our present facilities to a long chain of developments beginning with our ancestors who lived in the trees. Living in the trees required excellent hand-eye coordination, the ability to grasp and manipulate with the hands, and good binocular vision for depth perception. Later these ancestors began to move between the trees and the ground to forage for food. This new behavior required alertness to the threat of predators. Often surveying the area was accomplished by sitting up or standing for short periods to see over grasses and obstacles, especially at the edges of the forests.

Perhaps climate changes reduced forested areas, or maybe some of the forest dwellers were driven out into the savannahs by others. But whatever the reason, some of the pre-men began to encounter an entirely new set of conditions on the open grasslands. Here not only was an upright posture advantageous for increased mobility, it held decided survival benefits in that upright posture gave the pre-man a larger silhouette to bluff enemies and freed the hands to hold clubs and stones.

Conditions on the open savannah were severe and challenging. Competition for food was keen, requiring a high degree of flexibility and adaptability. Through trial and error, methods were developed for locating new sources and types of food. Conditions favored small, mobile, closely knit groups. The socialization of the young became more complex and required longer training periods. As social interaction became more complex, the rudiments of culture began to emerge.

But, while the line of Homo was the most socially advanced of the pre-men, its members were by no means the sole authors of culture. The Gracile Australopithecines, and to a lesser extent the Robustus variety, had the unmistakable marks of social beings even if their social structures were somewhat less developed.

In trying to understand why the pre-men might have found it necessary or advantageous to move away from the relatively safe haven of the trees and out onto the grasslands, several elegant and fascinating theories have been proposed. Naturalist Robert Ardrey has made a remarkable set of observations on animal behavior in relation to territory. Ardrey defines "territory" as:

...an area of space, whether on water or earth or air, which an animal or group of animals defends as an exclusive preserve. The word is also used to describe the inward compulsion in animate beings to possess and defend such a space. A territorial species of animals, therefore, is one in which all males, and sometimes females too, bear an inherent drive to gain and defend an exclusive property...In most but not all territorial species defense is directed only against fellow members of the kind. (33)

According to Ardrey, a territory works to protect food supplies, to disperse populations in relation to environmental resources, and to ensure the selection of superior mates through mock battles over real estate. He further states that the attachment of breeding couples to territory is the best assurance that the offspring will not be neglected. (34)

One of the most notable effects of a territory upon the behavior of a particular species is the increased energy which it lends to its possessors for its defense. Conversely, the same territory acts as an inhibiting factor to intruders. T.H. Huxley observed that when one group of a particular species was compressed into an increasingly smaller territory by adjacent competing groups of the same species, a corresponding counter-pressure was generated upon the repressed group to restore the territory to its original boundaries. The tighter the compression of a territory by similarly defended adjoining territories, the greater the competition for available resources and the greater the effort to protect the integrity of the borders. A territory thus has an elastic quality.

A territory, moreover, has degrees of importance attached to it according to how deeply it exists within a mass of similarly defended territories. The most valuable areas lie at the heart of a mass of territories and are reserved for the strongest, smartest and bravest of a species. Radiating out from this nucleus, the territories become less and less valuable. Conflicts, then, are the fiercest where contenders jockey for the prime real estate at the heart of the territories.

This innate drive to hold and defend territory extends to humans in many obvious ways. Ardrey proposes that one manifestation of this ancient territorial behavior is reflected in patriotism:

...the continuity of human evolution from the world of the animal to the world of man ensures that a human group in possession of a social territory will behave according to the universal laws of the territorial principle. What we call patriotism, in other words, is a calculable force which, released by a predictable situation, will animate man in a manner no different from other territorial species. (35)

Corroboration of these theories has been born out in the work of Jane Goodall and her colleagues at the Gombe Stream Reserve in Tanzania, where her studies of wild chimpanzees and other primates have helped to shed light on man's evolution. In a rare set of observations, Goodall and her associates recorded a primitive war between neighboring tribes of apes over a southern patch of territory in the park's Kahama region. This conflict began in 1974 and was finally ended with the death of the last male victim in 1977. (36)

Goodall believes that the phenomenon was initiated by four or five large males from a main northern tribe which mobilized when a number of chimpanzees broke away from the main tribe and established their own community to the south. This reduced the size of the northern tribe's territory, forcing them to resort to violence to reclaim it. Systematically the northern chimps isolated individuals, both males and females with young, from the main group. Each victim was savagely beaten, jabbed or bitten into a state of utter immobility and left to die. After the rebel tribe was effectively liquidated, the northern tribe was again observed to move freely through the southern portion of the reserve and was peaceful once more.

Sadly territorial "war" was not the only form of violence witnessed by Goodall for at least three members of the community were confirmed murderers! A female chimp named Passion, and later her offspring, Pom and Prof, carried out grisly rampages in which infant chimps were kidnapped, killed and eaten with relish. The reasons for Passion's infanticidal tendencies are unclear, but the actions certainly resemble a form of homicidal mania.

Chimpanzees are even known to hunt other wild animals for food, such as baby bush pigs, bush bucks and monkeys. Hunting parties are organized to stalk and kill the prey, and then primitive weapons are used to defend the catch from scavengers and other tribes.

It is in such behaviors that the qualities of our own nature are so graphically reflected, leaving little doubt about our common ancestry and the all-encompassing nature of evolution. Between the study of contemporary cousins, and the evidence of now extinct ones, it thus becomes easier to understand how early man developed.

Returning to our discussion of early man, we can perhaps begin to see that his move from the trees to the savannah materially increased his hunting instincts, and this hunting nature became pivotal to his cultural development. We know, for example, that he eventually learned to follow the seasonal migrations of deer and elk. For many of man's ancestors, the herd became his livelihood.

Loosely knit social groups evolved into highly integrated clans with specific roles and tasks assigned to each member. The strong hunted. The artful made tools. The women reared the young and prepared the food. Efficiency was crucial for survival. Durable clothing was made from the fur of man's victims. Bones became needles, awls and other tools; the sinew became thread; pelts became sleeping mats, and hides became shelters and moccasins.

Man spread far and wide, was challenged by nature and adapted. Even his physical structure changed to adapt to its surroundings. For example man's skin color became darker the closer he lived toward the equator and lighter the closer he moved toward the Earth's poles because of the skin pigment "melanin," which darkens the skin to protect the body from the increased dosage of harmful ultraviolet radiation found near the equator.

In colder climates man's physique became stockier, as with the Eskimos, to help conserve body heat, or he became taller and more slender in hotter climates, such as the African Maasai, for better heat dissipation. As he came to better control his environment, he also lost the need for the heavy covering of body hair of his ancestors.

Eventually man learned to use fire, first for defense from predators and for heat, then for cooking and for the hardening of tools. He was now becoming more than an instinctual animal driven by mere biological forces. He was becoming introspective. From his observations of the forces of nature, he began to make deductions about the nature and causes of such things as rain, the seasons, animal migrations, thunder and lightning, sickness, birth and death. Things that directly affected his existence, such as the stalking of an elk for food, took on special significance, for if the forces which controlled the animal could be influenced, then success of the hunt could be assured.

These mysterious forces eventually came under the jurisdiction of specially ordained medicine men who sought to manipulate or control them through acts of symbolic and sympathetic magic. For example fireside dances miming the killing of an elk were thought to produce the required spiritual dominance over a real animal, thus assuring tribal hunters would kill an elk on the hunt. To accomplish this, a pelt worn by the medicine man may have been symbolically speared by an aide to induce the desired kill. The pelt was thought to take on the "spirit" of the intended victim.

As time went on, the list of forces grew to many hundreds, both good and bad. The dead were thought to continue on as spirits for we know that various ancient tribes buried their dead not so much out of veneration for the dead body as to protect it from scavengers which could rob the soul from it. There was, of course, remorse for the loss of a loved one to the spirit world.

One of the most famous discoveries of an ancient burial site was made by archaeologist R.S. Soleki at the Shanidar Cave in the Zargos mountains of Iraq. According to Leakey, one summer day some sixty thousand years ago, a man was buried in a grave with bouquets of wild flowers arranged neatly about him. By studying the arrangements of the long lasting flower pollens, archaeologists have determined that he was lavishly arrayed with yarrow, corn flowers, St. Barnaby's thistle, groundsel, grape hycanthiths, woody horsetail and a type of mallow. The body itself was laid on a mat of carefully woven woody horsetail, clearly demonstrating that such elaborate ritual was already well established even by this very early time. (37)

The earliest forms of religion therefore employed mostly primitive magic. Man observed the forces of nature and tried to manipulate them to his own ends. To these impersonal forces were added the more personal "spirit" forces thought to reside in all living things, as with the native Americans, who saw the eagle as a kind of spiritual brother.

Man's need for some type of religious experience is demonstrated in the fact that no culture, existing or dead, has ever been found without religion in one form or another. It is man's innate compulsion to assign the ultimate responsibility for things which cannot be explained or understood to a superior order. Thus religion arose with man and evolved with him. It is a distinctly *man-made* institution. The spontaneous generation and evolutionary nature of religion is thus very helpful in explaining both the diversity of religions in general and the evolution of Judaism in particular.

The significant difference to be pointed out about the early Hebrew religion is that its birth and early roots reveal a curious mix of the aforementioned traditional religious beginnings and strong evidence of the proposed intervention of alien beings during their early history. The Starmen appear to have capitalized on the early Hebrew's innate compulsion for religious experience by allowing themselves to be made out as "gods" for the purpose of manipulating the primitive masses toward their own ends. The resulting "religion" was quite sophisticated, and its stature among the earth's religions today is testament to its formidable beginnings.

Meanwhile man's biological and cultural development up to this point had progressed at a relatively steady pace. We know this because from his tenuous beginnings in the *Ramipithecus* line nearly 13 million years ago to the late *Homo erectus* stage at about 100,000 years ago there was a gradual increase in the size of his brain from 400-450 cubic centimeters to a respectable 11-1200 centimeters.

Then an interesting thing happened about 65,000 years ago: The lineage of *Homo erectus* suddenly branched into two distinctly new variants—*Homo sapiens neanderthalensis* and shortly thereafter *Homo sapiens sapiens* (Cro-magnon man). Within just the last few thousand years, man's cranium grew to as large as 2,000 cubic centimeters with the average for the Cro-magnon line at about 1,400 cubic centimeters and the Neanderthals averaging slightly larger at 1,500 cubic centimeters. According to many experts, this is a huge increase in brain size considering the relatively short span of time in which it has occurred! Scientist-author Carl Sagan reports:

> *So far as I know, childbirth is generally painful in only one of the millions of species on earth; human beings. This must be the consequence of the recent and continuing increase in cranial volume. Modern men and women have braincases twice the volume of Homo habilis'. Childbirth is painful because the evolution of the human skull has been spectacularly fast and recent.* (38)

Moreover, there was a corresponding leap in the social sophistication and cognitive abilities, especially in Cro-magnon.

S. L. Washburn has suggested that the critical advantage of Cro-magnon over Neanderthal was, in fact, his use of language:

Traditions of stone-tool manufacture, as reflected in the rise of successive stone-tool industries, persisted for hundreds of thousands of years with little change. Then came the great acceleration of about 40,000 years ago. Men who were anatomically modern now dominated the scene. Primitive forms of man disappeared; there are not enough fossils to make it possible to decide whether the disappearance was by evolution, hybridization, or extinction. Then, in far less than 1 percent of the time that bipeds are present in the fossil record, came a technological revolution. Its fruits included entirely new and complex tools and weapons, the construction of shelters, the invention of boats, the addition of fish and shellfish to the human diet, deep-water voyages (to Australia for example), the peopling of the Arctic, the migration to the Americas and the proliferation of a lively variety of arts and a wide range of personal adornment. ...Agriculture and animal husbandry appeared at roughly the same time around the globe. ... What we see in the last 40,000 years of pre-history may have been triggered by the development of speech as we know it today. (39)

From Washburn's quote two additional points of interest emerge concerning the relationship between scientific views and Christian views on the evolution of man. First it is stated that the great acceleration of technological advance in man coincides with the advent of language. Interestingly in the New Testament, the Gospel of John states that all things for the Hebrews began with "the Word:"

In the beginning was the Word, and the Word was with God, and the Word was God. He was in the beginning with God; all things were made through him, and without him was not anything made that was made. In him was life, and the life was the light of men. The light shines in the darkness, and the darkness has not overcome it. (John 1:15)

Does this esoteric passage refer to the bestowing of speech upon the Hebrews as the stimulus for their religion? The author of John seems to say that everything is attributable to the "logos," which is Greek for both "word" and for "reason." The historic faith and reliance on the "word" or institution of God by the Hebrews cannot be overemphasized. It was everything to them. While God's word is now written in the

Bible, it was once delivered directly to the Hebrews, and it included every kind of instruction, including, in the beginning, speech.

A second thing to note about Washburn's quote is that with the rapid decline of the primitive Neanderthal man came the equally abrupt rise of the modern variant, Cro-magnon man, a stock completely different from the Neanderthals, which had preceeded them by 70,000 years or longer. Anatomically and culturally much more sophisticated, Cro-magnon spread far and wide across the globe from exactly where no one really knows. Some have speculated that it was from Europe, where he was originally discovered; others say Africa or the Middle East. One thing is sure: the differences between Neanderthal and Cro-magnon present an interesting anomaly for archaeologists. So different are the two, both intellectually and physically, that some archaeologists argue that Neanderthal to Cro-magnon is too big an evolutionary leap without an intermediate proto-man to bridge the gap:

In addition to stopping abruptly, the classic Neanderthaller is replaced with equal abruptness by people like ourselves (Cro-magnons). There is no blending, no gradual shading from one type to the other. It is as if modern man came storming in and dispossessed the Neanderthallers. (40)

To be sure there were variants of both kinds of beings throughout the world at one time, and some argue that the races did gradually blend and evolve to form the races we know today. But the astounding differences between the two variants, combined with at least two more facts -- the rise of civilization in the Middle East, and the rise of the Hebrew religion, also in the Middle East -- point to the amazing series of correlations upon which this book focuses. Let us now return to our original discussion of the Genesis passage in which God is said to have created man in His image.

The Intervention

The twenty-sixth verse of the creation reads:

1:26 Then God said, "Let us make man in our image, after our likeness; and let them have domination over the fish of the sea, and over the birds of the air, and over the cattle, and over all the earth, and over every creeping thing that creeps on the earth."

1:27 So God created man in his own image, in the image of God he created him; male and female he created them.

Several interesting things come together here. The Hebrew word for God as found in this portion of Genesis was originally *Elohim,* plural for *gods.* Subsequent editors have quietly removed this plural form in most of the Old Testament in favor of the singular form *El.* But the implication remains quite clear from the remaining passage above that whoever "created" man was a plurality!

It has been suggested by some theologians that the portion of the verse which reads, "Let us..." refers to God and his angels, or even to the Trinity; the Father, Son and Holy Spirit. It has been shown, however, that the Trinity was a relatively late concept in the Christian Church, and was never used by the Hebrews.

Angels, on the other hand, *are* present throughout the Bible, and figure quite prominently in the Old Testament stories. While they were not traditionally given credit for being "creators," God's use of angels to carry out his plan for the Hebrew race is well-documented. Therefore, it is suggested that the "angels" were starmen under the direction of a supreme leader, The Lord, who intervened in the natural evolution of man to produce a race of hybrids, thus initiating the Hebrew religion. Such a statement leads us to ask, "Where then, did the starmen come from?"

If we return to consider for a moment our scenario of the evolution of the universe, we are reminded that the laws which forged the stars, the planets and life on earth are universal. Not only could these laws have produced other intelligent life, they must have! Scientist-author Carl Sagan has said:

The number of advanced civilizations in the Milky Way Galaxy today depends on many factors, ranging from the number of planets per star to the likelihood of the origin of life. But once life has started in a relatively benign environment and billions of years of evolutionary time are available the expectation of many [scientists] is that intelligent beings would develop. (41)

If evolution *has* produced other intelligent beings before us, might they not already have ventured to this planet? What makes this idea so cogent is the fact of our own short history of space ventures, which are sure to increase dramatically in the near future. Ray Bradbury, the prominent science fiction writer, has a vision concerning our future in space, a vision which would not be unlike that of a people on any planet were they to occupy a position such as ours as we stand at the doorway to our space future:

As of the Apollo landings on the Moon, we have ensured our place in the universe. We can be immortal now. We can move out to the planets, the stars. We can colonize the Moon and Mars. We can move out to Alpha Centauri, and when we do that, our race will go on forever...and all the knowledge that we have developed in the past 15,000 to 20,000 years will go on with us. Shakespeare will survive and live with us in space. Christ will be there in space. (42)

When we have finally made space our permanent home, we will begin to move about freely through the universe, studying life forms, having wars, exploiting resources and, just possibly, lesser humanoids. Our advanced state of organization and knowledge would truly make us gods to any lesser beings. As Erich Von Daniken put it:

...although [the Starmen] can explain to the savages in their own language that no gods have landed, that no higher beings worthy of adoration have paid a visit, it will have no effect. Our primitive friends simply do not believe it. The space travelers come from the stars; they obviously have tremendous power and the ability to perform miracles. They must be gods! (43)

Barry H. Downing put the story of the Bible more succinctly: *"...the Bible argues in effect that the angels caused the religion of the Bible under God's direction."* (44) Although Downing and I have come to different conclusions concerning the results and purpose of the Starman visitation, we agree that a series of alien visitations did occur in man's dim history, that the visitations resulted in the development of the Hebrew religion, and that the story as told in the Bible has been completely distorted by subsequent misunderstanding and Church manipulation.

It is proposed, in summary, that man evolved naturally to a fairly modern state before being intervened upon both physically and culturally by alien Starmen to produce the Hebrew race and possibly others. In addition it appears that we have been continually monitored throughout the ages, since very early times when we were guided and instructed in such a way that it shaped a culture and inspired a major western religion.

This being said, there is yet one cornerstone to lay in the foundation upon which the Starman Intervention Theory rests. In order to understand more fully the breadth and the nature of the proposed theory, it will be instructive to briefly examine and compare evidence regarding the more recent Unidentified Flying Object (UFO) phenomenon in the hope that the reader will begin to see the links between the past and present phenomena and to provide references for future discussions.

Endnotes

15. Lance Morrow, "In the Beginning: God and Science," *Time,* Feb.5, 1979, pp. 148-50.

16. Editors, *Did Man Get Here by Evolution or by Creation?,* Watch Tower Bible and Tract Soc., New York, 1967, pp. 23-4.

17. Isaac Asimov, *The Universe,* Avon Books, New York, 1966, p. 47.

18. *Ibid.,* p. 221. (Some sources give a figure of 10 billion, others as high as 25 billion years as the age of the universe.)

19. *Ibid.,* pp. 79-80.

20. *Ibid.,* p. 130.

21. Peter W. Stoner, *Science Speaks,* Moody Press, Chicago, 1968.

22. Asimov, *op. cit.,* p. 136.

23. *Merit Student Encyclopedia,* MacMillan Ed. Corp., New York, 1978, p. 147.

Endnotes (continued)

24. "Evolution," *Scientific American,* a collection of articles on evolution, various authors, 1978, p. 35.

25. *Ibid.,* p. 49.

26. Catherine Jarman, *Evolution of Life,* Bantam Books, New York, 1972, p. 111.

27. *Merit, op. cit.,* p. 189.

28. *Ibid.,* p. 187.

29. Jarman, *op. cit.,* p. 115.

30. Michael H. Day, *Fossil Man,* Bantam Books, New York, 1970, p. 10.

31. Richard E. Leakey, *Origins,* E. P. Dutton, New York, 1977, p. 13.

32. Day, *op. cit.,* p. 18.

33. Robert Ardrey, *The Territorial Imperative,* Dell Publishing Co., New York, 1966, p. 3.

34. *Ibid.,* p. 87.

35. *Ibid.,* p. 213.

36. Jane Goodall, "Life and Death at Gombe," *National Geographic Magazine,* May, 1979, p. 592.

37. Leakey, *op. cit.,* p. 125.

38. Sagan, *op. cit.,* p. 93.

39. Washburn, *Scientific American, op. cit.,* p. 112.

40. "Early Man," *Life Nature Library Series,* Time-Life Books, New York, 1971, p. 126.

41. Sagan, *op. cit.,* p. 230.

Endnotes (continued)

42. Ray Bradbury, "God Is Alive in This Part of the Universe," *U.S. News & World Report,* June 25, 1979, p. 56.

43. Erich Von Daniken, *Chariots of the Gods,* Bantam Books, New York, 1969, p. 10.

44. Barry H. Downing, *The Bible and Flying Saucers,* Avon Books, New York, 1968, p. 15.

Exodus 24:9-11

Chapter V
MODERN UFOs

Any sufficiently advanced technology would appear indistinguishable from magic.

- Arthur C. Clark

My father once told me of an experience he had as a young man in San Manuel, Arizona, in 1957. He was a miner in those days, laboring long hours in the ore mines, and I was but two years old. One afternoon about four o'clock as he and several other men stood near the elevator preparing to descend for the night shift, someone pointed to the sky. There, near the horizon, was a very bright object hovering motionless. It appeared to be very large in relation to the surrounding terrain and very shiny. Then the mysterious object was seen to move across the horizon in a succession of rapid, measured movements. As it moved from place to place, it presented either a circular or a cylindrical shape, depending upon its orientation to the viewers. The incident was witnessed for approximately 20 minutes by most of that particular crew and others in the vicinity until the object suddenly shot off into the distance and disappeared over the horizon.

Dad told me that story when I was about nine years old and was becoming interested in UFOs. He makes no judgments about what he saw that day, and he has told few people of it, that I know of. But he did see something extraordinary; that I believe. The fact is many thousands of people have seen such phenomena in the skies and have wondered about their origins. But if the theory of evolution suggests the parallel developments of intelligent beings elsewhere in the universe, then logic would tell us that some of the more advanced races would have already conquered space and could be visiting this planet. The sheer numbers of earth-like planets in just our galactic neighborhood alone pose statistical probabilities which clearly elevate such an idea above the realm of mere speculation. Indeed, we should *expect* that some of the UFOs seen today are intelligently controlled space vehicles - the *same* vehicles which have been seen throughout history. Therefore the next logical step to understanding The Order is to examine briefly the UFO phenomenon, how it has been dealt with in the United States, and what its existence means in the light of history.

UFOs have long held a certain fascination for man. The mysterious objects which fly at incredible speeds, turn on a dime, appear in one spot and then disappear only to reappear in another, flash or glow or gleam in all colors of the rainbow, etc., seem to have defied every law of nature. Reported in every size and shape imaginable, by individuals and by groups, by every nationality, occupation, philosophical viewpoint, age group and generation, it has been a perpetual mystery.

Recent investigations have disclosed that about 80 per cent of these sightings can be explained in terms of natural phenomena, such as mistakenly identified flying aircraft, balloons, atmospheric conditions, birds flying in formation, swarming insects, planets, stars and hoaxes. There remains, however, 20 per cent of all sightings which cannot be so easily explained away, which defy all reasonable attempts to classify them into previously described catagories.

Such reports have been made under excellent viewing conditions, by competent citizens. Some of the most important and credible cases involved witnesses who were experts in the fields of astronomy, physics, meteorology, radar and photography; the list of observer occupations could fill a dictionary of careers. Many witnesses have avoided publicity, asking not to be named or photographed for fear of ridicule or loss of privacy. Some sightings have been independently confirmed by two or more witnesses, with the aid of instruments and sophisticated radar. Consider the following case from official Air Force Project Blue Book Files:

Rapid City Case

Just after dark on the night of August 12, 1953, the Air Defense Command radar station near Rapid City, South Dakota, received a call from an aircraft spotting post about 10 miles away reporting that observers could see an extremely bright light low on the horizon to the northeast. After the center was alerted, they studied the object for a few minutes. It appeared to be moving, but very slowly. The radar operators, aware of the fact that weather could sometimes affect their equipment, were careful to double check that the object was real; it was well-defined, solid and bright. They confirmed that the object was at 16,000 feet. Throughout this incident the spotter and the radar center maintained phone contact to compare the radar and visual observations.

Suddenly the spotter relayed that the object was picking up speed and heading toward the southwest. Two men from the radar station ran outside to look and yelled that they too could see a large blue-white object moving toward Rapid City. Now, with the three independent sets of observers, there was no doubt that they were all looking at the same object. According to the report, the UFO made a wide sweep around the city and returned to its original position in the sky.

The Warrant Officer in charge and his staff of radar operators, who had combat radar experience from Korea to Europe, declared they had never in their entire careers seen anything like it before. At this point the radar officer decided to vector an F-84 operating in the vicinity to the area of the object. The pilot quickly arrived on the scene, to within approximately three miles of the UFO, and reported visual contact. But suddenly it began to move again. This was confirmed by the spotter, the air traffic controller and the radar center.

Once it began to move, the UFO picked up speed fast and started to climb, heading north, but the F-84 was right on its tail. The pilot would notice that the light was getting brighter, and he'd call the controller to tell him about it. But the controller's answer would always be the same, "Roger we can see it on the scope." (45)

Apparently the closest the object would allow the pilot to approach was three miles, for as the pilot maneuvered his plane and tried again to close in, the object responded by pulling away to maintain the three-mile buffer. The pilot chased the UFO to about 120 miles north of the base before a low fuel indication forced him to break off and head for home. Then a frightening thing happened:

Both the UFO and the F-84 had gone off the scope, but in a few minutes the jet was back on, heading for home. Then 10 or 15 miles behind it was the UFO target also coming back. (46)

The F-84 came in to land. On the ground the fighter squadron which had been listening to the action on its radio, asked to scramble another F-84 to intercept the object. The controller gave the OK, and a second plane took off toward the "light." The pilot saw the object right away and closed in:

Again the light began to climb out, this time more to the northeast. The pilot also began to climb, and before long the light, which at first had been 30 degrees above his horizontal line of sight, was now below him. He nosed the '84 down to pick up speed, but it was the same old story - as soon as he'd get within three miles of the UFO, it would put on a burst of speed and stay out ahead. (47)

The pilot began to question himself even though he clearly saw the object and confirmed it with the ground station. He extinguished his cockpit lights to be sure he wasn't getting a reflection on his canopy. No change. Perhaps a light from the ground was causing the problem? To check he rolled the plane, but the light remained fixed. It wasn't a star; the object clearly moved in relation to three prominent nearby stars. He then reasoned that if it was a real object, the plane's radar should pick it up too. It did. The object was real and solid.

At this time the pilot asked to break contact and return to base. What else could he do? He was a seasoned flier, having engaged enemy MD-109s, FW-190s and ME-262s over Germany and MIG-15s over Korea, but this object really shook him. If it was indeed an aircraft, he was badly outclassed. This intruder was clearly identified by six independent sets of witnesses as making intelligent movements in protected airspace, but our planes, the best in the world at the time, were useless to intercept it!

Shortly thereafter the object left the area, heading toward Fargo, North Dakota. A few minutes later the Rapid City base received a call that spotter posts a few miles west of Fargo had spotted a large blue-white light traveling at high speed.

Thousands of similar reports have been filed by professionals such as these, competent observers, well-trained and experienced in their respective fields yet confounded by this alien presence. These reports provide evidence which has been carefully analyzed by several independent sources, including the Air Force. The conclusion: UFOs are intelligently controlled vehicles which fly in logical patterns and are probably extraterrestrial. Today we call them UFOs or Unidentified Flying Objects. In the times of Abraham and Jesus, they were called "swift clouds," "pillars of fire," "fiery chariots" and "cherubim."

Several authors have pointed out that ancient literature, from the Vedas of India to Shakespeare, holds a wealth of documented sightings if only we saw them for what they really were. Some material is rather vague while some is quite explicit. Jacques Vallee points out a particularly interesting account from the *Annales of Laurissenses* for the year 776 A.D., which reads:

Now when the Saxons perceived things were not going in their favor, they began to erect scaffolding from which they could bravely storm the castle itself. But God is good as well as just. He overcame their valour, and on the same day they prepared an assault against the Christians, who lived in the castle, of whom many still live to this very day, say they beheld the likeness of two large shields reddish in colour in motion above the church, and when the pagans who were outside saw this sign, they were at once thrown into confusion and terrified with great fear they began to flee from the castle. [sic.] (48)

In his book *Gods and Spacemen in the Ancient East,* W. Raymond Drake has assembled substantial material from ancient literature which seems to document this extraterrestrial presence throughout history. Among the richest of the ancient writings, according to Drake, are the Sanskrit texts, which have been dated from 1,500 to 4,000 B.C. (49) This literature is thus some of the oldest known to man, and some experts assert that it may even have influenced the comparatively newer biblical literature.

Sanskrit scholars like the erudite Dr. Max Muller, agree that the Vedas are far more ancient than Homer and form the real theogony of the Aryan race; in comparison the cosmogony and theogony of Hesiod and Genesis appear crude images of the Vedic sublimity. (50)

For example in the ancient *Ramayana,* composed of some 24,000 verses, are to be found wonderfully explicit verses describing heavenly beings which for a time resided on the earth, teaching man many things, having wars and riding their celestial cars:

Sailing o'er the cloudless ether Rama's Pushpa chariot came

And ten-thousand jocund voices shouted Rama's joyous name
Silver swans by Rama's bidding soft descended from the air
And on the earth the chariot lighted car of flowers divinely fair. (51)

From Book One, Chapter IV of the same work:

And the Gods in cloud-borne chariots came to view the scene so fair
Bright Adityas in their splendour, Maruts in the moving car
Bright immortals gaily crowding viewed the scene surpassing fair
Heavenly blossoms soft descending with a perfume filled the air
Bright celestial cars in concourse sailed upon the cloudless sky
Drum and flute and harp and tabor sounded deep and sounded high

From the *Samsaptakabadha Parva,* page 58, we read of the motions of these celestial cars in battle:

O, King, in the battle between the Gods and the Asuras in the days of old, [the celestial car] displayed a circular, forward, backward and diverse other kinds of motion... (52)

In a passage from the *Budhasvamin Brihat Katha Shlokassamgraha* as translated from a twelfth century manuscript by Felix Lacote in 1908, we discover the landing of a flying machine very similar to an account by the biblical prophet Ezekiel:

In the morning in the cloudless sky, the ascetics heard a rolling, which filled the atmosphere. "What is it?" they asked the Aerial Being. "It is the noise of the drums of the aerial voyagers, as they are enclosed in the bosom of the chariots, one hears them like a rolling of thunder. Here comes our Master, the King of Kings of the Vidyaharas with the storm of drums rumbling by the paths of heaven. Look!" Like a troupe of clouds that the rainbow illuminates, the lightnings and the cranes reclothed in golden splendour, filling all the spaces of the firmament, a troupe of chariots...appeared to the ascetics in the distance arriving from the sky. The chariots descended, that of Shakravatan halted at the door of the hermitage, the others in the gorges, on the backs and tops of mountains. The chariot of the supreme King of the Vidyaharas had the form of the lotus-flower ornamented with twenty-six petals, which were ruby. He himself was standing in the middle of the pericarp, formed by an emerald and on the petals poised his wives marvelously attired. (53)

These passages from the ancient Indian literature represent but a few of the copious references to such vehicles, beings, their weapons and other advanced technologies. The ancient literature is full of references to interactions between extra-terrestrials and man. But even before the advent of writing, there were visitations by strange beings from space. There are substantial materials, such as art and artifacts, which graphically demonstrate that the primitives in our dim history knew of or had contacts with extra terrestrials. Good examples of UFO drawings are to be found in the Cro-magnon caves of France and Spain, which date back to some 15,000 years ago! Upon those walls are to be found accurate renditions of bison, horses, reindeer and mammoths, which clearly attest to the artistic abilities of those peoples. And yet there is more:

...along with the animals, scattered among the various caves are a variety of incomprehensible signs and shapes which students of prehistory have for a long time tended to ignore, or at best treat as magic or fertility symbols...Many of the oval and disk shapes depicted by those Paleolithic artists agree, feature for feature, with descriptions given by twentieth-century man of things he has seen in the sky. The caves appear to contain a complete catalog of contemporary designs. (54)

More recent sightings before 1947 rarely attracted attention: Generally considered oddities and reported in scientific journals, ships' logs and police reports in a rather perfunctory manner, they now take on a new significance. Consider this example from an 1887 edition of *L'Astronomie,* which was labeled "ball lighting:"

In the North Atlantic ocean a new case of ball lighting, one of those so strange and still so inexplicable effects, has been observed. On November 12, 1887, at midnight, near Cape Race, a huge ball of fire appeared slowly emerging from the ocean to an altitude of sixteen or seventeen meters. This sphere started moving against the wind and stopped close to the ship from which it was observed. Then it rushed away in the sky and disappeared in the southeast. The whole observation had lasted five minutes. (55)

Another interesting sighting occurred aboard the *USS Supply* on February 28, 1904, by Lieutenant Frank H. Schonfield at 6:10 A.M.:

They appeared to be traveling in echelon, and so continued as long as in sight...The largest meteor had an apparent area of about six suns. It was egg-shaped, the sharper end forward... When the meteors rose there was no change in relative position... That they did come below the clouds and soar instead of continuing their southeasterly course is also equally certain... The meteors were in sight over two minutes and were carefully observed by three people whose accounts agree as to details. (56)

Note that in each case the objects in question were referred to in terms of things that were already familiar to the observers. This is thought to have occurred throughout history up to the late 1940 s. Then conditions changed drastically, and all perception of what UFOs could really be was irreversibly altered.

The popular term "flying saucer" was coined in 1947 by a businessman named Kenneth Arnold, who reported spotting nine very shiny disk-like objects while flying his private plane near Mt. Rainier in Washington. According to Arnold, the objects resembled two saucers placed lip to lip and then skipped across a body of water like stones across a pond. Their flight was ordered and very fast, much faster than that of an airplane. The incident became one of the most widely reported and most thoroughly documented cases in the history of the UFO phenomenon. It was a landmark UFO case, not only because of the broad public attention it drew, but also because it was one of the first cases to be thoroughly investigated and scrutinized by the U.S. Air Force.

The Arnold sighting ignited the nation's curiosity, and the term "flying saucer" quickly became a household word. Today nearly all dictionaries include a definition for the phenomenon as *"a saucer-shaped illuminated object often reportedly seen in the skies."* (57) The controversy which was to rage over the next several decades, however, found the term drifting into the realm of satire or lampoon, and it is now largely avoided by serious students of extraterrestrial theories. The term Unidentified Flying Object (UFO) is the official Air Force designation and more objectively identifies the phenomenon we are addressing.

The key to the magnitude of interest in UFOs after Arnold's historic sighting may be tied to the level of technical sophistication which we have attained. Before the advent of high-powered jet aircraft and rockets, few people recognized the potential correlation between UFOs and our own ability to travel in space.

But the time interval from the late forties through the sixties generated a tremendous burst of awareness in the possibility of life from other planets visiting the Earth. Stories of space travel and invading aliens, such as Jules Verne's *From the Earth to the Moon* and Orson Welles' *War of the Worlds,* set the stage, giving people the proper mental orientation. Then the new post World War II technologies, such as jets and rockets, added the credibility.

Once the concept of breaking the bounds of the earth was realized, the weight of the correlation hit home: UFOs could be vehicles with space travelers like us! The result was sensational. The eyes of the world turned to the heavens. Reactions varied: some were awed, and still others were disgusted with the notion, as UFOs became one of the most hotly debated subjects for nearly three decades.

Government Investigations

No small part of the controversy was due to the Air Force's sudden keen interest in these numerous objects which intruded into our airspace with such impunity. The mere fact that an agency of the United States government would take the time and energy to investigate and evaluate UFOs was more than enough to generate widespread interest in the matter.

Underlying the roar of controversy, however, was the Air Force's apparent inadequate handling of the whole affair. Unfortunately there was a storm of controversy which was later published that clearly suggested the Air Force had a split-level policy on its UFO investigation. On one level the Air Force collected data from individuals while claiming there was no evidence to support the idea that UFOs were intelligently controlled alien vehicles. On the other hand, the Air Force and other agencies secretly held there might be some merit to the idea that UFOs *were* alien vehicles.

In addition the investigations themselves suffered from some rather dubious evaluation procedures. Especially during the early days of the investigations, most reporting routines, it is claimed, lacked consistent data collection and analysis procedures. But worst of all, some of the high-ranking directors and personnel on the projects were so openly biased about UFOs that internal dissent and conflict among committees and staffs often nullified any hope for objectivity. It became a very emotional issue instead of a scientific inquiry. This was carefully screened from public view, however, and has only recently come to light.

One of the foremost members of the Air Force projects, who served as a scientific consultant on the astronomical aspects of the UFOs for nearly 20 years, was Dr. J. Allen Hynek. His scathing indictment of the Air Force's handling of UFOs is detailed in his book *The Hynek UFO Report.* In it he states:

> *[Once] the decision was made that UFOs had to be figments of the imagination, the Air Force policy on UFOs never changed direction. "It can't be, therefore it isn't" became the guiding principle, and anyone associated with [Project] Blue Book, from the Director on down, learned to follow suit or else... When it became clear to me and others within the project that the official Pentagon policy was to debunk UFO sightings, intelligence analysts and investigators alike did their best to come up with "common sense" explanations for each new UFO report.* (58)

But this wasn't the secret opinion of the Defense Department, for there were some expressing grave concern that this phenomenon was no freak of nature or circumstance. The evolution of the various "projects" which dealt with UFOs began as a simple investigation, but as public awareness grew, so did pressures to find satisfactory answers to what was occurring. The response was to develop the split-level system: one of public pacification and data collection, and the other a secret evaluation of the still unresolved enigma.

At approximately the same time Arnold sighted his famous disks, the Army Air Force was undergoing a massive reorganization with the objective of becoming a service branch separate from the Army. As public interest in UFOs swept the country, teams of Army Air Force intelligence investigators were dispatched to interview witnesses. All data was analyzed by the Air Material Command (AMC) at Wright-Patterson Air Force Base near Dayton, Ohio.

In 1947 a report was sent to the Commanding General of the Army Air Forces. AMC findings were that the reported phenomenon were *real,* not visions or hoaxes, that the objects were solid, metallic appearing, elliptical or circular in shape, flat on the bottom and domed on the top. The following memorandum, presented by R. E. Fowler, describes the AMC's opinion concerning the UFOs in the investigation. It is dated 23 September 1947, and is addressed to the Commanding General, Army Air Forces, Washington, 25, D.C., Attention: Brigadier General George Shulgen:

As requested by AC/AS-2 there is presented below the considered opinion of this command concerning the so-called "flying discs." This opinion is based on interrogation report data furnished by AC/AS-2 and preliminary studies by personnel of T-2 and Aircraft Laboratory, Engineering Division T-3. This opinion was arrived at in a conference between personnel from the Air Institute of Technology, Intelligence T-2, Office Chief of Engineering Division, and Aircraft Power Plant and Propeller Laboratories of Engineering Division T-3.

It is the opinion that the phenomenon is something real and not visionary or fictitious. There are objects probably approximately the shape of a disc, of such appreciable size as to appear to be as large as a man-made aircraft...The reported operating characteristics such as extreme rates of climb, maneuverability...and actions which must be considered evasive when sighted or contacted by friendly aircraft and radar, lend belief to the possibility that some of the objects are controlled either manually, automatically or remotely. The apparent common description of the objects is as follows:

(1) Metallic or light reflecting surface
(2) Absence of [vapor] trail except in a few instances when the object apparently was operating under high performance conditions
(3) Circular or elliptical in shape, flat on the bottom and domed on top
(4) Several reports of well-kept formation flights varying from three to nine objects
(5) Normally no associated sound except in three instances a substantial rumbling roar was noted
(6) Level flight speeds normally above 300 knots are estimated

The letter is signed by N. F. Twining, Lt. General, USA, Commanding. (59)

On December 30, 1947, the Commanding General of the Army Air Forces initiated a new investigation called Project Sign. It was given a level 2A priority and ordered to conduct a thorough investigation into the phenomena. Responsibility for Sign fell to the Air Technical Intelligence Center (ATIC), which was an integral part of the newly reorganized Air Force. According to Edward J. Ruppelt, former chief of a later UFO project, ATIC concluded from its investigation that these objects were, again, not figments of the public's imagination, Russian super weapons, natural phenomena or hoaxes, but *interplanetary vehicles.* But the public was apparently given a completely contradictory story, for less than a year later, on December 27, 1948, the Air Force made the following public announcement:

The Air Force has discontinued its special project of investigating and evaluating reported flying saucers. The reports are a result of misrepresentation of various conventional objects, a mild form of mass hysteria, and hoaxes. Continuance of the Project is unwarranted. [Official Air Force release] (60)

Unknown to the public at the time, however, the project did not close, but changed names and reopened secretly in February, 1949, under a new code name - Project Grudge! The project surfaced again in 1952 when public pressure prompted the Air Force to do something about a massive UFO sighting wave which was sweeping the world. Thus in March, 1952, the project was again renamed and brought out into the open, this time as Project Blue Book. By that time then Project Chief Edward J. Ruppelt noted that UFO reports were flooding into his office at the rate of 20 to 30 per day.

Project Blue Book was the longest lived and best known of the Air Force investigations. Much of the credit for its success and credibility must go to Mr. Ruppelt, whose organizational skills and insistence on accurate, objective reporting helped boost the old Project Grudge from a poorly run, rather disorganized department within a department to a responsible organization worthy of the public trust. New reporting procedures were instituted, forms and classification criteria were standardized and improved, pilots and crews were briefed on current knowledge, and outside groups were contacted to prepare statistical computer studies and do other research. In short, Ruppelt was largely responsible for making the Air Force's investigative effort legitimate again. To prevent a relapse into some of the problems of the past, he imposed a strict ban on openly passing judgments about the project and even fired some of his staff for breaking the rule.

Shortly afterward, on January 12, 1953, the Central Intelligence Agency (CIA) convened what has now become known as the Robertson Panel, after its chairman, the late, renowned scientist H. P. Robertson of the California Institute of Technology. The CIA had become increasingly aware of the ramifications of the Blue Book Project's findings and was concerned that if UFOs weren't in and of themselves dangerous to the security of the nation, then perhaps the Russians might try to exploit the general confusion about them to somehow cause us to believe that their attacking bombers were only UFOs, etc.

Although the CIA sponsored the conference, and it was given the appearance of an objective inquiry into the situation, there is evidence that it had a hidden agenda. The supposedly objective panel, some of whose members are suspected of being affiliated with the CIA, studied only a handful of the literally thousands of reports in Project Blue Book's files. Detailed reports which experts spent hundreds of hours investigating and compiling were given only a cursory review. In all only 12 hours total time was spent reviewing all of the Air Force's evidence over a three-day period, casting suspicion on how serious the panel was about weighing the evidence. Still unaware of the situation, Ruppelt had what he thought would make the decisive argument in favor of more detailed study:

The next item on the agenda ...was a review of a very hot and very highly controversial study. It was based on the idea that Major Dewey Fournet and I had talked about several months before - an analysis of the motions of the reported UFOs in an attempt to determine whether they were intelligently controlled. The study was hot because it wasn't official and the reason it wasn't official was because it was so hot. It concluded that the UFOs were interplanetary spaceships...

The study covered several hundred of our most detailed reports. By a very critical process of elimination, based on the motion of the reported UFOs, Fournet told the panel how his and any previous analysis by Project Blue Book had been disregarded and how those reports [of phenomena] that could have been caused by ...astronomical bodies, etc., were sifted out. This sifting took quite a toll, and the study ended up with only ten or twenty reports that fell into the "unknown" category. Since such critical methods of evaluation had been used, these few reports proved beyond a doubt that UFOs were intelligently controlled by persons with brains equal to or far surpassing ours. [sic] (61)

Ruppelt had guessed correctly: Most of the members who reviewed the reports recommended that the Blue Book program be expanded fourfold and that immediate procedures and scientific studies be instituted to better analyze and monitor UFO activities. They advocated keeping the public informed and instituting checks and balances to reduce bureaucratic bungles or cover-ups. Not all of the panel's recommendations were in accord with the wishes of the CIA, however, and CIA representatives on the panel recommended that there be a *debunking* of UFO reports:

The "debunking" aim would result in reduction in public interest in "flying saucers" which today invokes a strong psychological reaction. This education could be accomplished by mass media such as television, motion pictures and popular articles. The basis of such education would be actual case histories which had been puzzling at first, but were later explained. As in the case of conjuring tricks, there is much less stimulation if the secret is known. (62)

The Robertson Panel thus marks another major shift away from the Air Force's sometimes open and objective handling of its UFO investigations. From this point on, the function of Blue Book regressed to a data collection and public relations agency, and the real investigative effort shifted elsewhere. The panel's blatant statement of intent to mislead the public underscores the government's dual treatment of the UFO phenomenon: military secrecy on one hand and public pacification on the other. For several years thereafter, the Air Force was fairly successful toward this end. Between 1964 and 1967, however, there occurred another large wave of UFO sightings which so aroused the public concern that the Congress of the United States was obliged to hold an open hearing on the matter:

In response to demands from angry constituents, Weston Vivian, Democratic Congressman from Ann Arbor, and House Republican Minority Leader Gerald Ford, called for a congressional investigation into the Air Force's handling of UFO reports. (63)

Again a special investigative panel was formed. At the requests of J. Allen Hynek and others, a panel of outside experts was brought together to scientifically analyze UFO reports and present their findings to the Air Force. And again there was difficulty from the start. The contract was awarded to the University of Colorado with a directive specifying that Dr. E. U. Condon, professor of Physics, be solely in charge.

Dr. Condon, a former director of the National Bureau of Standards, had an impressive scientific background and a long association with military research projects. According to Blum (64), Condon was a pioneer of experimental physics in the U.S. and was instrumental in the development of radar, the atomic bomb and the nose cone and heat shields for the Mercury and Gemini space projects. He was also apparently an outspoken critic of governmental abuses concerning such things as radioactivity levels and public safety; so much so, in fact, that his Top Secret security clearance was twice revoked without any official justification. Mysteriously, about a year before his appointment to the directorship of the Colorado Project, as it was called, he was again awarded his Top Secret clearance.

From the beginning, the committee found itself embroiled in internal dissent on the UFO issue. Dr. Condon and his chief administrative assistant, the late Dr. Robert Low, were both rather openly opposed to the notion that UFOs could be extraterrestrial and allowed their personal biases to set the tone for the entire project, despite protests by subordinates that they were ignoring important evidence and the principle of objectivity.

The contract between the University of Colorado and the Air Force carried with it an expressed mandate that the work be conducted under conditions of strictest objectivity, but Condon's attitude about the whole affair was more in accord with the CIA's objective of debunking the whole thing, and it appears that Dr. Condon's role was just that. Speaking before a chapter of Sigma Xi, an honorary scientific fraternity on January 25, 1967, Condon said of the project:

"It is my inclination right now to recommend that the government get out of this business. My attitude right now is that there's nothing to it." With a smile he added, "But I'm not supposed to reach a conclqsion for another year..." (65)

Another disruption occurred when two members of the committee were fired for revealing to the public a memo written by the Administrative Assistant, Dr. Low, which states in part:

The trick would be, I think, to describe the project so that, to the public, it would appear a totally objective study but, to the scientific community, would present an image of a group of non-believers trying their best to be objective but having an almost zero expectation of finding a [flying] saucer. (66)

Reactions to the committee's final report were mixed, but the sober conclusion among many authorities is that the project was a farce, and the voluminous report totaling 965 pages was a hatchet job. There was a storm of controversy that Condon was ignoring the fact that nearly 30 percent of the cases which he was investigating fell into the "unknown" category and that he was openly making conclusions contrary to some of the other panel members', whose comments urged further study of the matter. But even if the committee was more objective, there was still the problem of their methods. According to J. Lorenzen, Director of the Aerial Phenomenon Research Organization, an independent research group:

The Condon Committee formally closed its doors in 1968 and published its report. Their investigative methods had been sadly lacking ... and what we saw of their files indicated that they had no standardized report form, no set method of investigations and generally the project was haphazardly conducted. (67)

Despite the controversy and discord among the panel members, Condon's and Low's views that further studies of UFOs "probably cannot be justified in the expectation that science will be advanced thereby" (68) was what the Pentagon wanted to hear. On the basis of this recommendation, the Air Force again announced that it was closing down its UFO investigation, and on December 17, 1969, Project Blue Book was formally closed.

The Condon Committee's findings could not have been timed more precisely, for the annual symposium of the prestigious American Association for the Advancement of Sciences (AAAS) was to devote that year's seminars in Boston to the subject of UFOs. Although the symposium went ahead as scheduled, the combined effects of the much publicized Condon Committee's report and the subsequent closing of Project Blue Book greatly diminished its impact.

Despite this double blow to the scientific recognition of UFOs as a legitimate phenomenon in need of further study, the American Institute of Astronomics and Aeronautics later did its own study of the Condon Committee's report and made this official statement:

We find it difficult to ignore the small residue of well-documented but unexplainable cases that form the hard core of the UFO controversy. (69)

The conclusion to be drawn from this information is that the Air Force has allowed terribly obvious corruption of its public investigative projects. The tone and direction of its efforts became grossly misleading to the public, even to the extent of leaving out facts on reports, fabricating data to make reports conform, and convening panels to discredit legitimate observations.

Is the Air Force finished with UFOs? Although there is no longer a formal public UFO project, the responsibility for handling UFO reports now falls under the jurisdiction of the Aerospace Defense Command (ADC). Evidence suggests that there is a continued interest in UFOs, but the Defense Department is not currently involved in any public relations on the matter.

The current official stand on UFOs may delay but will not extinguish the scientific recognition of the possibility of extraterrestrial visitors. J. Allen Hynek, Edward J. Ruppelt, Raymond Fowler, Jacques Vallee, David M. Jacobs, the Lorenzens, the Blums and others have produced voluminous data and case studies to show beyond a reasonable doubt that UFOs are indeed intelligently controlled extraterrestrial vehicles. In his closing statement to his book *Challenge to Science: The UFO Enigma,* Vallee says:

The "UFO phenomenon," with all its disturbing aspects, presents a most unwelcome challenge to the physical and philosophical conceptions of the universe painstakingly formed in the course of many centuries of civilization on this planet. But it is there; we cannot forever refuse to study it. And it could well be that, in the final analysis, our own existence will be dependent upon the sincerity with which we conduct this research. (70)

The fact that the former project director for the Air Force's Project Blue Book should come out to reveal the true story from the military side, and then Dr. Hynek, one of the Air Force's former ardent UFO debunkers, who later reversed his judgments and came out to tell the story from the civilian side, has effectively circumvented the official cloak of secrecy. If the extraterrestrial origins theory for UFOs is correct, we shall eventually have our "close encounter of the third kind." It is only a matter of time.

UFO Categories

Before moving on to the inspection of contact cases, there is yet one brief point to address about UFOs, namely the classes or types of UFOs. The voluminous evidence of UFOs on file in the various organizations and agencies reflects the enormous diversity of reported UFO design configurations, i.e. ovals, cigar shapes, spheres, cylinders and saucer shapes, etc. In the systematic study of UFOs, it becomes quite evident that there are at least two and probably more types or classes of ships. The *reasons* for these differences are not important - they may be different types of vehicles, differences in interpretation by observers, or from different planets. We are primarily interested in two classes: mother ships, or those UFOs reported as launching other small UFOs, and smaller single UFOs. Both types have been reported around the world in various shapes, but the form of most interest to the thesis of this book is the cylinder shape, variously described as the "cigar" or "pillar" shaped UFOs, which correspond remarkably well to the phenomena reported in the Bible. Three such "pillar" type UFO reports are presented here for later comparison.

The first report comes from Vernon, France, where a cigar-shaped UFO was sighted by M. Bernard Miserey on August 23, 1954:

Just before 1 a.m., M. Bernard Miserey, a businessman of Vernon, arrived home and put his car away. As he came out of the garage, on the south bank of the Seine [River], he was surprised to see a pale light illuminating the town, which had been in complete darkness a little while before. Looking at the sky, he saw a huge, silent, motionless luminous mass, apparently suspended above the north bank of the river some three hundred yards away. It could have been compared to a gigantic cigar standing on end...

"I had been watching this amazing spectacle for a couple of minutes," M. Miserey later reported, "when suddenly from the bottom of the cigar came an object like a horizontal disc, which

dropped at first in a free fall, then slowed, and suddenly swayed and dived horizontally across the river toward me, becoming very luminous. For a very short time I could see the disc full-face; it was surrounded by a halo of brilliant light. A few minutes after it had disappeared behind me, going southwest at prodigious speed, a similar object came from the cigar and went through the same maneuvers. A third object came, then a fourth. There was a long interval, and finally, a fifth disc detached itself from the cigar, which was still motionless. This last disc dropped much lower than the earlier ones, to the level of the new bridge, where it remained still for an instant, swaying slightly. At that time I could see very clearly its circular form... and the glowing halo surrounding it. After a few seconds pause, it wobbled like the last four, and took off like a flash toward the north... During this time the luminosity of the cigar had faded, and... had sunk into the darkness."

When M. Miserey described his nocturnal vision the next day, the police informed him that two policemen making their rounds at about 1 a.m. had also observed the phenomenon, as had an Army engineer who had been driving along Route N-181 southeast of Vernon. (71)

That sighting compares favorably with a sighting in Ipswich, Massachusetts on September 17, 1966:

Mrs. David Rogers woke up at 4:45 a.m. with a coughing spell. On the way to get a drink of water, she noticed a light shining through a window that overlooked a lonely shoreline known as Crane's Beach. The Ipswich River and the sea meet on this narrow bay bordered by the Plum Island reservation on one side and the huge forested vacated Castle Hill estate on the other. On a typical summer's day this area swarms with boats, bathers and fishermen, but soon after Labor Day, the miles of sand dunes play host only to sea gulls, sandpipers and nature lovers. Mrs. Rogers glanced at the lighted window... nothing should be out on the beach now she thought. Curious, she walked over to the window to see where the light was coming from. She took one startled look and ran to wake up her husband to come and see. There, hovering at the water's edge over the beach, a bare half mile away, was a huge elongated, cigar-shaped object standing on end. It glowed with a dim, golden, pulsating white light, which at times brightened enough so that the beach sand could be seen at its base. At times it would tilt slowly to about sixty degrees from the vertical and then move back to the upright position. Finding it hard to believe his eyes,

David quickly stepped out onto an open-window porch to get a better look, with his wife trailing nervously behind him. It was no optical illusion. It was really there! The night was clear with no noticeable wind. A sound similar to breaking surf came from the object's direction. Suddenly, two small bright lights appeared moving just above the beach toward the glowing upright cylinder. Astonished, the couple watched them move "with a skipping motion, sometimes fast, sometimes slowly," and merge with the top of the cylinder! Almost instantly, three similar lights emerged from the top of the cylinder and sped off low over the ocean. One object appeared to momentarily approach within a quarter mile of them. Fascinated, they watched the smaller objects enter and leave the larger cigar-shaped object several times. Then, the object dimmed and just disappeared from sight. (72)

On September 14, 1966, 250 miles from Paris, France, a similar object was observed by hundreds of people from several villages during daylight hours:

It was about five in the afternoon. Emerging from the thick layer that looked like a storm coming up, we saw a luminous blue-violet mist, of a regular shape something like a cigar or a carrot. Actually, the object came out of the layer of clouds in an almost horizontal position, slightly tilted toward the ground and pointing forward like a submerging submarine.

This luminous cloud appeared rigid. Whenever it moved, its movements had no connection with the movements of the clouds, and it moved all of a piece, as if it were actually some gigantic machine surrounded by a mist. It came down rather fast from the ceiling of clouds to an altitude which we thought was perhaps a half-mile above us. Then it stopped, and the point rose quickly until the object was in a vertical position, where it became motionless.

During this time the dark clouds went on scudding across the sky, dimly lighted from underneath by the violet luminosity of the object. It was an extraordinary sight, and we watched it intently. All over the countryside other farmers had also dropped their tools and were staring up at the sky like us.

All at once white smoke exactly like a vapor trail came from the lower end of the cloud. At first it pointed to the ground but finally rose up to describe around the vertical object an ascending spiral. While the rear of the trail was dissolving in the air and being carried off by the wind, the source of the trail went up to the very top of the vertical object and then started to come down again, turning the other direction. Only then, after the smoke trail had vanished entirely, could we see the object that was sowing it - a little metallic disk, reflecting in its rapid movements, flashes of light from the huge vertical object. The little disk then stopped turning around the luminous cloud and went down toward the ground again, this time moving away. For quite a few minutes we could see it flying low over the valley, darting here and there at great speed, sometimes speeding up, then stopping for a few seconds, then going on again, flying in every direction between villages that were four miles apart. Finally, when it was almost a mile from the vertical object it made a final dash toward it at headlong speed and disappeared like a shooting star into the lower part, where it had first come out. Perhaps a minute later the carrot leaned over as it began to move, accelerated and disappeared into the clouds in the distance. The whole thing lasted about a half an hour. (73)

We now have the proper orientation to look at the words of the Bible and understand, perhaps, what Moses was saying when we read of the "pillar of cloud by day and pillar of fire by night" which led the Israelites into the wilderness. Let us now consider some of the cases of contacts with these alien beings to see how they support the Starman Intervention Theory.

Endnotes

45. Edward J. Ruppelt, *The Report on Unidentified Flying Objects,* Ace Books, Inc., New York, 1956, p. 306.

46. *Ibid.,* p. 307.

47. *Ibid.,* pp. 306-7.

48. Jacques Vallee, *Anatomy of a Phenomenon,* Ace Books, Inc., New York, 1965, pp. 21-22.

49. W. Raymond Drake, *Gods and Spacemen in the Ancient East,* Signet Books, New York, pp. 31, 36.

50. *Ibid.,* p. 31.

51. *Ibid.,* p. 38.

52. *Ibid.,* p. 44.

53. *Ibid.,* p. 53.

54. Ralph and Judy Blum, *Beyond Earth,* Bantam Books, New York, 1974, p. 39. Drawings also on p. 40.

55. Vallee, *op. cit.,* p. 32.

56. Blum, *op. cit.,* p. 59.

57. "Flying Saucers," *Webster's Handy College Dictionary,* Signet Books, New York, 1972, p. 186.

58. J. Allen Hynek, *The Hynek UFO Report,* Dell Publishers, New York, 1977, pp. 33-4.

59. Raymond E. Fowler, *UFOs: Interplanetary Visitors,* Bantam Books, New York, 1974, p. 221, (E. U. Condon, *Scientific Study of UFOs,* p. 894-5.)

60. *Ibid.,* p. 229.

61. Ruppelt, *op. cit.,* pp. 286-7.

62. Fowler, *op. cit.,* p. 231, (E. U. Condon, *Scientific Study of UFOs,* p. 915.)

63. Blum, *op. cit.,* pp. 157-8.

64. Blum, *op. cit.,* p. 159.

65. *Ibid.,* pp. 160-1.

66. Fowler, *op. cit.,* p. 156.

67. Coral and Jim Lorenzen, *Encounters with UFO Occupants,* Berkley Medallion Books, New York, 1976, pp. 4-5.

68. Hynek, *op. cit.,* pp. 282-3.

69. Fowler, *op. cit.,* p. 158, (*UFO, An Appraisal of the Problem,* AIAA, Nov., 1970.)

70. Jacques and Janine Vallee, *Challenge to Science: The UFO Enigma,* Ballantine Books, New York, 1966, 3rd Printing, p. 234.

71. Fowler, *op. cit.,* pp. 204-5.

72. This report is from the personal files of Raymond Fowler, Report #66-42, *UFOs: Interplanetary Visitors,* p. 203.

73. Vallee, *op. cit.,* pp. 107-8.

Chapter VI
CONTACTS

Do not neglect to show hospitality to strangers, for thereby some have entertained angels unawares.

- Hebrews 13:2

The question has been raised, "If extraterrestrials truly ply our skies, then why haven't they contacted us?" The answer is they probably have. Contacts with spacemen have occurred constantly throughout history and continue to be reported today. Old and ancient literature holds a wealth of accounts of angels, demons, sylphs, gods and assorted other beings who came from the skies and contacted man. It appears, however, that the more recent contacts have been restricted to very low-key individual relations. Speculation as to why a superior race of spacemen might operate in this manner would be ludicrous, but the evidence remains that they come. The understanding of the story behind the Bible can be enhanced through a brief inspection of some more recent reports of such contacts.

What word we get of contacts today is the sensational stuff of gossip magazines and talk shows. Hence the viewing of such incidents objectively is an elusive goal. If a poor soul is unfortunate enough to run into some space beings, he/she is likely to suffer not from the contact, but from public scrutiny following the incident. Still contemporary media coverage and public airing of contacts carry less penalty today than during the middle ages. During the reign of Charlemagne, for example, there seem to have been so many contacts with these demons, or sylphs, as they were sometimes called, that both Charlemagne and his successor, Louis the Debonair, imposed "grievous penalties upon all these supposed Tyrants of the Air." (74)

An interesting account has been found regarding Agobard, the Archbishop of Lyons, France (c. 840 A.D.), who relates that upon hearing a commotion, he ran outside to find four people being assembled in preparation to be stoned to death because they were foolish enough to admit they had ridden aboard an "aerial ship." Fortunately Agobard convinced the mob that such a thing was quite impossible:

Out of the number of those whose blind folly was deep enough to allow them to believe these things possible, I saw several exhibiting in a certain concourse of people, four persons in bonds - three men and a woman who they said had fallen from these same ships; after keeping them for some days in captivity they had brought them before the assembled multitude, as we have said, in our presence to be stoned. But truth prevailed. (75)

If Agobard believed in such things himself, he did not admit it, but other prominent historical figures, including mathematician Jerome Cardan (1501-1576) and Swiss physician Aureolus Paracelsus (1493-1541) clearly felt that such beings did exist. An interesting corroboration of Agobard's account is to be found in A. H. Clough's introduction to Plutarch's *Parallel Lives*:

One day, among other instances, it chanced at Lyons that three men and a woman were seen descending from these aerial ships. The entire city gathered about them, crying out that they were magicians and were sent by Grimaldus, Duke of Benventum, Charlemagne's enemy, to destroy the French harvests. In vain the four innocents sought to vindicate themselves by saying that they were their own countryfolk, and had been carried away a short time since by miraculous men who had shown them unheard-of marvels, and had desired to give them account of what they had seen. The frenzied populace paid no heed to their defense, and were on the point of casting them into the fire, when the worthy Agobard, Bishop of Lyons, who having been a monk in that city had acquired considerable authority there, came running at the noise, and having heard the accusations of the people and the defense of the accused, gravely pronounced that both one and the other were false. That it was not true that these men had fallen from the sky, and that what they said they had seen there was impossible.

The people believed what their good father Agobard said rather than their own eyes, were pacified, set at liberty the four Ambassadors of the Selphs, and received with wonder the book which Agobard wrote to confirm the judgment which he had pronounced. Thus, the testimony of these four witnesses was rendered vain. [sic.] (76)

In a treatise entitled *Why These Beings Appear to Us,* Paracelsus brilliantly proposes that these beings are angels:

Everything God creates manifests itself to man sooner or later. Sometimes God confronts him with the devil and the spirits in order to convince him of their existence. From the top of Heaven, he also sends the angels, his servants. Thus these beings appear to us, not in order to stay among us or become allied with us, but in order for us to become able to understand them. These apparitions are scarce, to tell the truth. But why should it be otherwise? Is it not enough for one of us to see an angel, in order for all of us to believe in other angels? (77)

It seems natural to assume that actual contacts are less often reported than the simple observations of funny lights in the sky. For one thing, contacts are a quantum perceptual leap in the public's eye toward the confirmation of extraterrestrial life. For another thing, there has been just enough negative publicity generated by some contact cases to keep a healthy skepticism alive on the subject. And furthermore, contacts often involve such strange, even bizarre images and experiences that it is difficult for both victims and the public to comprehend and acknowledge them.

Contemporary reports of contacts seem divided among the spectrum of experience, from the obvious hoaxers to the apparently legitimate contacts by extraterrestrials. In the broad grey middle area are the UFO cultists, whose beliefs in UFOs and spacemen actually form a variant form of religion centered around the visitations of benign space beings who bring messages. The range of beliefs regarding the significance of UFOs has produced such colorful personalities as the late George Adamski, whose close-up photos purporting to show alien space craft were circulated during the fifties and sixties, and "Bo and Peep," whose mysterious visitor led a small band of followers to a new enlightenment.

Carl Sagan holds to the theory that belief in UFOs and contacts with spacemen without the concrete scientific evidence to back it up is really another form of religious experience based upon man's aforementioned innate need for some type of religious experience to bridge the gap from the known to the unknown in an uncertain world and is therefore only a variation on the hundreds of other varieties of religious interpretation that exist.

While it is true enough that such cults exist, just as do hoaxers, it is also clear that some who report contacts with alien beings have nothing to do with either. While the cultists' claims that aliens regularly contact them may or may not be valid, their experiences really provide no tangible solutions or answers to scientific inquiry except that it is in keeping with the thesis of this book that people who witness things they do not understand continue to form a religion around them just as they always have.

While Sagan does not appear to believe in UFOs as real alien spaceships, other notable scientists, such as Dr. J. Allen Hynek, continue their scientific research into the phenomenon using the scientific method and specialists from many fields in their quest for an answer. It is interesting to note that many astronomers, Sagan included, believe that life may exist on other planets, but they remain to be convinced that intelligent life has visited Earth until more "concrete" evidence is available. So while hoaxers and cultists provoke public skepticism, behind the scenes the subject is very much alive.

Entire books have been published on the following contact/abduction cases, and those presented represent but a sample of many good accounts. While they may appear quite bizarre, their validity should not be judged on that aspect alone. Consider also the circumstances and the evident moral characters of the people who have reported the events.

There are chilling threads of continuity running through these cases, which were widely separated in time and space. Are these incidents to be treated separately, or do they paint a bigger picture when taken together? In no way can we be certain that every one of these experiences happened exactly the way we read them now or even that they happened at all. Most of the following stories have created some controversy.

In the end, it is up to the reader to sort these experiences out in his or her own mind. Whatever the case, such records of contacts can be traced into the dim reaches of history. The following recent examples are presented as preparation to considering another series of contacts during biblical times.

Walton

Travis Walton was one of seven loggers working on November 5, 1975, in the rugged Sitgreaves National Forest a few miles outside Heber, a tiny town in the heart of Arizona on Highway 260 at the juncture with Highway 277. Their job was to clear a fire break to help check the movements of forest fires in the dry season.

The long workday finally ended about dusk, and so they packed up their gear and made their way back down the winding dirt access road in an old International crew-cab truck. Suddenly one of the men pointed out a strange light spiking through the trees ahead of them. As they approached the area, the tree line gave way to a fantastic sight - a luminous object hovering motionless just below the tree tops.

Walton later described the object as having an approximate diameter of 15-20 feet, being 8-10 feet high and giving off a soft golden glow. He compared it to a flattened disk, like two pie plates placed lip to lip, with sharply defined edges and topped with a small white dome. He thought the object was divided into translucent panels, which glowed, by a dull, grey frame.

The men filed out of the truck and moved slowly toward the object, stopping a good distance away to observe. According to Walton, he recalls hearing a faint mechanical sound emanating from the object from his observation point of what he estimates as about 30 yards away. After a few moments of observation, Walton was the one first overcome more with curiosity than fear and began to ease toward the object. While the others chastised him to stay back, he eased toward a slash pile of cut trees and brush to get a better look. Then something began to happen:

> *Suddenly, I was startled by a powerful, thunderous swell in the volume of the vibrations from the craft. I jumped at the sound, which was similar to that of a multitude of turbine generators starting up. I saw the saucer start wobbling on its axis with a quickening motion, in a pattern like the erratic spin of a slowly moving top. The same side stayed toward me while it wobbled.* (78)

First he ducked. Then he made a move to get out of there. But as he stood up, he was hit by a tremendously bright blue-green beam which shot out from the underside of the craft. He was knocked violently to the ground and struck unconscious.

His companions panicked. They thought they were being attacked and fled the scene toward town to get help. Meanwhile Travis was apparently taken aboard the ship, for when his friends returned with the authorities, he was nowhere to be found. An extensive search of the area produced no sign of him.

When the six co-workers first reported the incident to the local sheriff, many of the townspeople thought the men had conspired to murder Walton and had disposed of the body. Because of the seriousness of the allegations, all agreed to undergo polygraph examinations to lay to rest the question of the murder and to help substantiate their incredible story of the UFO. Each was given three separate tests by Cy Gilson, the Arizona Department of Public Safety Officer, who later stated: *"These polygraph examinations prove that these five men did see some object they believe to be a UFO..."* (79)

The sixth co-worker, Allen Dalis, tested "inconclusively," meaning no accurate stress evaluation could be obtained because of his level of anxiety over the whole affair. The examiner's opinion was that, *"Allen was just too agitated to be tested at all. Even if his charts were readable and showed foul play, he couldn't have committed a crime and made up a story without involving five other men whose tests corroborate the report."* (80)

Five days later, on November 11, Travis was deposited in a semi-conscious state near Highway 260 a few miles outside Heber. He recalls feeling rather sickly, exhausted and disoriented. He did not remember what had happened after the blue-green beam had struck him. He made his way into town and spent most of his time dodging the media, undergoing medical and polygraph tests and being extensively interrogated.

But the most interesting aspect of the Walton case came out under hypnosis. At first Travis had little or no recall of his five absent days. However, with the help of Dr. James Harder, Director of Research for the Aerial Phenomenon Research Organization (APRO), a carefully conducted investigation and hypnotic regression was performed to help clear up details of the fantastic story and to ascertain, among other things, the credibility of the witness.

While under hypnosis Travis recalled that when he first came out of his unconscious state, he realized he was inside the craft and was undergoing what appeared to be a physical examination by three human-like beings. Among the most prominent features of the beings, which both terrified and intrigued Walton, were the small-framed bodies with large heads and enormous eyes. Of the eyes Walton said:

The only feature that was not underdeveloped were those incredible eyes! Those glistening brown orbs had irises twice the size of a normal human eye. The iris, or colored part of the eye, was nearly an inch in diameter! The iris was so large that even parts of the pupils were hidden by the lids, giving the eyes a certain cat-like appearance. There was very little of the white part of the eye showing. They had no lashes, no eyebrows... I could not stand to look at their eyes, but I found my own kept returning to look into theirs. It was impossible to avoid their compelling stare... In spite of my terror, I felt there was also something gentle and familiar about them. In fact, their overall look was disturbingly like that of a human fetus! (81)

With pale, almost translucent skin, no visible hair, thin-lipped mouths which uttered no sound, small crinkled ear lobes that lay close to their heads, and miniature noses with small oval nostrils, they had all the features of men with variations in proportion, size, color and texture.

Slowly coming back into consciousness, Walton saw only foreign images. Above where he lay was a peculiar device which he realized the beings were using to examine him. With a rush of adrenaline, Walton jumped off the examination table screaming the threats and warnings of a trapped animal as he eased to the back of a wedge-shaped room. Then, just as he prepared for a confrontation, the beings raised their hands in an apparent gesture of deference and left the room.

He waited for a few moments and then moved tensely toward the entrance to a curved passageway. No sign of the beings - only a dimly lit corridor. He then made his way down the passage to the entrance to what he said looked like a piloting room with a control chair. As he approached the chair, the room darkened into a planetarium with a panorama of stars and galaxies projected around him. After a few more moments, he sat in the chair and began to press some buttons and move levers. He noted that when he moved one lever, the universe spun in various directions.

Suddenly a different, very human-like being appeared in the room and approached Walton with a small contingent of the other smaller beings. This second being was human-like in every respect. He wore a jump suit and a bubble-sphere helmet. Walton tried to talk to this being with no response and was escorted out of the room, down the hall, and apparently *out* of the vehicle and into what appeared to be a hangar with other vehicles. They were met by yet another couple of similar beings, one of which appeared to be a woman.

Walton thought of resisting but decided to go along with the program until he could figure out what to do. He was helpless until he could figure out where he was, what this was all about, and if it was real, although it certainly seemed real. He was led to another table, where he was promptly lifted up and put to sleep by some self-contained anesthetizing device weilded by the female. There was no time to resist.

The next thing Walton remembered was waking up and seeing a glimpse of what appeared to be the same vehicle he had seen before, as it flashed off into the night sky just west of Heber. The whole thing suddenly felt like a dream, like a distant event that was fading with the passing of each waking moment. Yet there was an eerie feeling that something quite real and terrifying had occurred.

When first made known, this story caused quite a sensation, and several books were written about it, including *Abducted* (Pocket Books) by Jim and Coral Lorenzen, *The Walton Experience* (Berkley) by Travis Walton himself, and *Ultimate Encounter* (Pocket), a book by Bill Barry.

Numerous aspects of this case resemble those of other reported abductions and contacts. Perhaps more of these stories might be told, but for fear of ridicule or other more damaging sanctions, many people have chosen to keep their bizarre experiences a secret. Some who have adamantly refused to believe their senses have tried to bury their experiences in their subconscious for weeks, months or even years. Then the nightmares and the self-doubts about their mental health begin to frighten or disturb them enough to seek professional help.

The men involved in the Walton case suffered scorn and ridicule, as have others who dared to report their experiences. In the Walton case, the incident was reported at a time of high stress for Walton's companions because Walton had disappeared, and their concern for his welfare was just enough to override their anxiety about revealing how it happened. While some of the locals were quick to accuse the logging crew of foul play, the accusations were put to rest only after extensive polygraph examinations and Walton's eventual return.

Was Walton really abducted by an alien space crew? Could it be a perverted hoax by some bored, small-town loggers looking to cause a sensation, as some have suggested? Quite possibly. Only time or Travis Walton will tell. But this case does have an interesting corroboration.

Moody

On the night of August 12, 1975, just three months prior to Walton's encounter, Air Force Sergeant Charles Moody was involved in a very similar incident near Holloman Air Force Base in Alamogordo, New Mexico. After his ordeal, he too felt sickly and exhausted for reasons he could not explain. He also felt that strange sensation that something odd, different, inexplicable had occurred, yet he too could not remember anything specific. But the thing which perplexed him most was the fact that he was somehow missing over an hour of time during the encounter for which he could not account. To add to his anxiety, he was developing a rash over his lower body and back and was still in some pain several days after the incident.

Moody did not report his experience to the Air Force for fear of losing his job or being transferred or some other undesirable repercussion. Instead he decided to write a letter to *The Official UFO Magazine* asking for help but requesting confidentiality. On September 2, field investigators were dispatched to check Moody's story.

Moody, at 32, was a thirteen-year veteran flight mechanic on Air Force cargo planes with a high security clearance. He was a family man whose wife was a nurse with two fine children and a nice home in Alamogordo. To investigators he seemed to be a solid witness. He was in good health, had no mental problems, had 20/20 vision, was a light drinker and took no drugs. He stated he had never seen a UFO before and did not believe in such things until this experience.

The night of the incident, Moody got off work at about 11:30 p.m., drove home, changed clothes and watched television until about 12:30. Still feeling wide awake, and having heard there was to be a meteor shower visible over the desert about 1:00 a.m., he drove to just outside town, to the edge of the White Sands Missile Range to watch the sky for a while. It was a clear night, and the desert sky was full of stars. Just as predicted, he began to see some wonderful meteor activity. Over the next half-hour, he remembers observing eight or nine bright meteors streak across the sky. But he also saw something he hadn't counted on.

At about 1:30 a.m., as he was sitting on the fender of his car, he spotted an object very similar to the one seen by Walton. According to Moody:

> *...a dull metallic object... seemed to just drop out of the sky and start to hover with a wobbling motion approximately one hundred feet in front of me, and approximately ten to fifteen feet off the ground. I was very frightened. The object was moving very slowly toward my car.* (82)

As the fifty-foot craft loomed closer, Moody jumped off the fender and got into his car to drive away, but the electrical system was completely dead - no lights, no starter. Scared almost to the point of panic but with nowhere to run, he sat frozen in his car and watched as a window appeared on the craft and through which humanoid forms could be made out. They were staring at him. He then remembers a tremendous increase in the noise level, a high-pitched sound like a dentist's drill, before a feeling of numbness came over his body. He described the feeling as "like floating on a cloud." This was accompanied by the loss of fear which had gripped him only seconds before. The next thing Moody remembers is the object lifting off very fast and disappearing into the night.

After the object left, the car started easily, and Moody headed for home. Then he received another shock when he glanced at his watch - it was 2:45 a.m.! When he had looked at his watch only moments earlier, or so he thought, it was only 1:20. Where had the "missing" hour and twenty-five minutes gone? Upon arriving home he told his wife of the incident and noted that the wall clock read 3:00. Although he looked pale and sick to his wife and felt he might need medical help, he deferred going to the military doctors immediately for fear of their reaction to his symptoms. The next day he wrote his letter to *The Official UFO Magazine* asking for help.

The usual process of investigating a case such as this involves dealing with the psychological and emotional trauma from the incident. Moody, who had previous commando training and claimed to fear no man, was visibly shaken by his experience. Although he could remember some of what had happened, there existed within him a feeling that just below his conscious grasp, like a dream he couldn't quite recall, was the answer to his fear.

A thorough investigation was conducted, including a visit to the site of the incident. Although the area showed no signs of a UFO landing, Moody's story seemed sound, as did his background, home life and reputation. The investigative team concluded that Moody was a credible witness and the event was interesting, but unfortunately, since he couldn't remember key parts of the encounter during the "lost" hour and twenty-five minutes, the story was considered too sketchy to be of great importance. Moody was asked to keep in touch if further details came to light so that documentation of his story could be completed. But the significance of this story did not become apparent until the APRO field team began to investigate the Walton case in mid-November.

On November 15, more than two months after the Moody investigation, and with the Walton investigation well under way, a new dimension to the Moody case was just coming to light. As time passed, Moody began to recall glimpses of his experience during what must have been the missing time period and began to write them down. Finally, in a personal letter to the directors of APRO, Moody recounted what he could remember in some detail. In it he described beings of almost identical proportions to those in the Walton case:

The beings were about five feet tall and very much like us, except their heads were larger and had no hair. Their eyes were larger than ours, ears very small, nose small, and the mouth had very thin lips. I would say their weight was maybe between one-hundred ten and one-hundred thirty pounds. There was speech, but their lips did not move... There were no names said, but they knew who I was and called me by my proper name, Charles, and did not use my nickname, Chuck. It was like they could read my mind, and I believe they did, because the elder, or leader, would speak sometimes before I would ask something. (83)

Moody's experience was different from Walton's in some respects, but various key points of the cases do coincide remarkably - the beings' mannerisms, appearance, outer garments, the interior design of the ship, etc.

Apparently Moody put up an initial struggle while being abducted, as did Walton. Both were also treated in some manner by the beings for adverse effects from it, though Moody doesn't remember a struggle, only of being told by the beings that there was one and a feeling of discomfort on some parts of his body.

Similarities between this and other cases are also to be found in the mental anesthesia effect caused by some unseen force which produced a numbness or "floating feeling," a sedated state often reportedly accompanied by a temporary loss of consciousness or ability of the body to respond to normal commands from the brain.

And, almost always, there is a mental or telepathic suggestion to forget the entire event. Eventually, however, the victims seem to undergo strange stresses. Vivid dreams of a bizarre nature or some emotional problem associated with the occurrence causes them to override their fear of persecution or ridicule to seek some type of professional help, some way to talk to a competent, impartial listener, to seek assurances that they are not losing their minds, and that some strange, alien event has indeed intervened in their lives.

Andreasson

Another fascinating, utterly bizarre abduction occurred on January 25, 1967, to Betty Andreasson, but it was nearly a decade later before she and her daughter, Becky, could relive the event under hypnosis to reveal the mystery.

For ten years Betty lived with a hazy memory of some strange and frightening event. Sometimes vivid segments would resurface as flashbacks. Her experience, like many others, was like a recurring distant nightmare that was somehow very real and troubling. Just why she decided to tell her story after so long she can't say, but as the urge became stronger, as her fears gave way to curiosity, she began to look for a way to get her story told.

She found that way in 1975, in an August edition of her local newspaper, which ran an article describing the Center for UFO Studies and its director, Dr. J. Allen Hynek. Taking pen in hand, she sat down and wrote a letter describing what she could remember of her experiences. When it was received by the Center, however, it was merely filed for future action.

When the case was finally followed up some months later, it was decided to have Betty and her daughter undergo some hypno-regression therapy because her story was still quite sketchy for lack of recall. The therapy was conducted by Harold J. Edelstein, director of the New England Institute of Hypnosis. The investigation which followed was lengthy and extensive:

> *...it has become probably the best documented case of its kind to date, the subject of an intensive twelve-month investigation conducted for the Center for UFO Studies (CUFOS) that involved, among other things, the recording of large quantities of testimony given under hypnosis, extensive lie-detector testing of witnesses, detailed analysis of corroborative circumstantial evidence, careful character checks, exhaustive comparison with other [contact cases], and much more... The result of this investigation filled three volumes of a 528-page confidential report.* (84)

Again an entire book was written on the case, entitled *The Andreasson Affair* by Raymond E. Fowler, and only a brief sketch will be related here. Betty Andreasson's children had just been bathed and fed and were in the living room watching *Bozo the Clown* on early evening television. Betty was in the kitchen finishing up her few remaining chores in preparation of visiting her husband, who had been hospitalized after an auto accident a few weeks earlier.

Suddenly the lights began to flicker and then went out. The house completely darkened, causing the frightened children to rush into the kitchen calling for their mother. At about the same time, the family saw a strange pink light come shining through the kitchen window, getting brighter, pulsating. Sensing something was wrong, Betty ordered the children into the living room. There was a terrible stillness in the house. Betty's father, who was staying at the Andreasson home, rushed into the kitchen to find out what was going on. When he saw the lights and looked out the window, he gazed in dismay upon an alien sight:

These creatures that I saw through the window of Betty's house were just like Halloween freaks. I thought they had put on a funny kind of headdress imitating a moonman. It was funny the way they jumped one after the other - just like grasshoppers. When they saw me looking at them, they stopped... The one in front looked at me and I felt kind of queer. That's all I knew...(85)

Apparently the beings froze everyone in the house except for Betty. After a few moments, the lights in the house came back on just as several beings, she can't remember how many, came into the kitchen *through* the door. They appeared to materialize right through the door for Betty recalls under hypnosis that it was not opened. Grasping for an explanation, she tried to understand what was happening:

I'm thinking they must be angels, because Jesus was able to walk through doors and walls and walk on water. Must be angels ...And the scriptures keep coming to my mind where it says, "Entertain the stranger, for it may be angels unaware." (86)

Betty, a devout Christian (although not a fanatic), seems to have drawn upon her fundamentalist beliefs to try to reconcile the situation with her knowledge. The beings advanced, one after the other, in a fluid leap-frog fashion or hopping motion as they approached Betty. She describes them as having large wrap-around cat-like eyes, holes for noses, small lipless slits for mouths. Their bodies were small with grey "clay-like" skin. Their heads were large in proportion to their bodies. They wore dark blue form-fitting suits with blue diagonal banners across their chests. On each sleeve was an emblem resembling a bird with outstretched wings. Their hands appeared gloved and were large and fleshy with three fingers each. When they communicated, they did so by telepathy. Note again that many of these features agree, item for item, with those of other encounters.

They entered the living room where the children sat motionless while Betty showed the leader, who was substantially taller than the other beings, a Bible. Betty recalls that she was not afraid, but the strange sequence of events kept her mentally off balance. For example the leader reproduced the Bible, as if by magic, and handed each subordinate a duplicate, which they each skimmed very rapidly.

Meanwhile, Becky Andreasson, who was at the time eleven years old, began to awaken from her trance-like state. Under hypnosis she too was able to describe the beings in some detail, and the fact that she was at first at ease with the thought of their presence but became frightened as they looked in her direction. She was especially distressed by their large eyes, which she described as like "large marbles." At that instant one of the beings noticed that Becky was not fully sedated, and she was quickly returned to her trance-like state. Betty also recalled under hypnosis the feeling of being sedated, though not as heavily as the other members of her family, and she too was impressed by the eyes:

> *And their eyes are so funny. One minute they are light, and now they've got a black ball in it. And they move straight, especially that left eye...* (87)

Betty asked why they came to her and what they wanted. They replied they needed her help and asked her to follow them. She later realized, when she tried to resist, that she really had no choice. The beings had come for a specific purpose, and their requests, it seems, were merely to ease the anxiety and fear of the experience. But at this time Betty did not feel threatened in any way and so followed them outside and into a strange craft which was shrouded in a dense patch of fog in the back yard.

The object is described as oval in shape with a raised portion in the middle. It stood on struts which adjusted to the slope of Betty's back yard and is described as being a metallic "silver-gold" color. Many of the details of the ship's interior are also revealed. Inside the ship the light seemed to permeate from everywhere. The furnishings were sparse and simple, and the doors opened and closed swiftly, leaving no visible seams. Such descriptions among reliable abduction cases are common.

Another common denominator between this incident and other abduction cases is that Betty was given a physical examination and a kind of "operation" was performed on her while on board the ship. According to Betty, the examination room was a simply furnished chamber in the shape of a dome. While lying on some type of examining table in the middle of the room with light coming from every wall, she described the event while under hypnosis, some ten years later, as if it were actually happening again. At times she became chillingly emotional. She did not want the examination and became very frightened. She described how the beings inserted a long flexible needle up her nostril and into her head. When she experienced pain, the leader placed his hand on her forehead and took the pain away through some sort of hypnotic suggestion or telepathy.

Another long bendable needle was inserted into her navel. Betty could hear the beings communicate with each other. Something was wrong, they were saying; there were parts of her body missing. And indeed there were, for after the birth of her last child, Betty underwent a hysterectomy. The most startling aspect of this whole episode was the revelation that these needles were designed to implant and extract tiny B-B sized spheres which enabled a machine on the ship to probe the body without opening it up! One studded pellet was apparently missing because of Betty's hysterectomy, and the beings were complaining because they couldn't "see" the lower half of her body!

Some unseen force held Betty down on the table so that all she could see was that allowed by the movement of her eyes in their sockets. At this point the beings decided to perform another test with what seemed to be a scanning device which lowered from the ceiling to her abdomen and did not hurt as did the needles. During the examination the aliens communicated constantly with her, reassuring her and placing a hand on her forehead whenever she experienced discomfort, which then abated.

This examination is remarkably like the one undergone by Betty Hill during her abductiontin 1961, and those in the Hickson case in 1973 and several others. The time duration for the entire event is not certain. Betty and her family were put to bed, and when they awoke, they appeared to remember nothing. Upon Betty's return the night before, it was learned under hypnosis that the leader put his hands on her shoulders and instructed her to forget what she had gone through, and she was given a message:

He says that he has had others here...Many others have locked within their minds, secrets...He is locking within my mind certain secrets...They will be revealed when the time is right... (88)

Repeated attempts under hypnosis could not get Betty to reveal what "secret" information she had locked in her mind. She either could not regress deeply enough in the hypnosis to retrieve the information or could not override the earlier hypnotic suggestion by the beings to keep the information secret.

A revelation no less awesome in its implications is the fact that Betty must have had the tiny pellets embedded in her long before the incident because during the examination the pellets were *removed.* Betty had been visited before, which she confirmed under hypnosis! The story ends on an ominous note: Betty could be contacted at any time for they always knew where she was. Interestingly the Moody case closely parallels in this respect:

The elder, or the leader, then put his hands on the sides of my head and told me that it was time for them to leave, and he asked me not to remember what had been said, or what I had seen, for at least two weeks... I asked [the elder] if we would meet again, and he said yes, in a short time. And he then told me to visit a doctor soon... I then asked why they had talked to me, and why I was taken on board their craft. He said only that "in time you will understand." (89)

Some aspects of the Andreasson case are quite bizarre and some have no parallels. For example, Betty was taken on a trip to a remarkable alien place which she describes in detail and, unfortunately, is beyond the scope of this book. Her experience is decidedly religious in overtone, but the author and investigative teams who spent many months on the case have speculated that perhaps the beings sensed Betty was a deeply religious woman and utilized or exploited her mental orientation to manipulate her. Perhaps she concocted a mixture of real and imagined events, or perhaps it didn't occur at all.

That would be hard to believe, however. Given the careful research and documentation, and the hypno-regressions by competent staff and investigators, it is clear it would have been very difficult for anyone to tell the same story over and over without bungling some part of it. Moreover, a person under hypnosis, it has been pointed out, simply does not lie. Betty

Andreasson, it appears, despite the formidable challenges to belief in her amazing story, was just telling what she knew.

There are many other excellent abduction cases which have been carefully researched and documented as well, including the case of Officer Herbert Shirmer, a policeman from Ashland, Nebraska, who on December 3, 1967, a mere eleven months after the Andreasson incident, had a contact with UFO occupants who gave him some "important information" which he was not to reveal until the "proper time." Charles Hickson and Calvin Parker, who were abducted from a river bank while fishing in Pascagoula, Mississippi on October 11, 1973, were also given physical examinations. Hickson recalls that he was "examined" by a round device suspended from the ceiling of an alien ship, which looked like an eye and was passed over the length of his body.

The list is long and growing, and the accounts are often quite bizarre, but one need not dismiss them for their strangeness. All of the cases related here have been the subject of the strictest scrutiny. The utmost care was taken to ensure the integrity, accuracy and mental fitness of the victims. Those who have dared to relate their stories have usually done so out of a need to understand some problem associated with their experiences, and though every one of these cases may have been the result of a hoax or a dream, given the characters of the witnesses and the volume of the circumstantial evidence, that remains a dubious alternative. Of the recent abductions, however, there remains one case which must be told, and that is the amazing case of Barney and Betty Hill.

Hill

It is appropriate that we consider one final case - the abduction of Barney and Betty Hill, not only because it is one of the most credible cases yet reported, but also because it is one of the most detailed, most scrutinized and most well-known. It is a landmark case. The Hills received widespread exposure from their claimed contact with the occupants of a UFO some five years after the event, when they were finally compelled to have the facts of their experience published due to some unfortunate and unauthorized publicity which sensationalized something the Hills wanted to keep to themselves.

Barney was an assistant dispatcher at the Boston post office and his wife Betty was a social worker for the state of New Hampshire. Both Barney and Betty also worked for the Portsmouth National Association for the Advancement of Colored People (NAACP), Barney as chairman, Betty as assistant secretary and coordinator of community projects. On top of that, Barney was a member of the State Advisory Board for the United States Civil Rights Commission and the Board of Directors for the Rockingham County Poverty Program. Betty also served as United Nations envoy to the Unitarian-Universalist Church to which they both belonged in Portsmouth.

Both Barney and Betty were extremely reluctant to discuss the episode with any but their closest friends initially for fear of ridicule or unfavorable reactions from associates. The incident was obviously controversial in nature and they could not risk being branded as eccentrics with their important civil rights work, to which they were both deeply committed.

With a rash of local UFO sightings getting high visibility in the national press, a reporter for a Boston newspaper got hold of the Hill story and published details without consulting the Hills, which greatly upset them. When it was learned that Mr. John G. Fuller was in the area investigating the phenomena, the Hills sought him out to tell their story with all the details. Fuller was selected because his recent research into another string of over sixty UFO cases at Exeter, New Hampshire brought him by chance to their area. Fuller's familiarity with the subject of UFOs, their characteristics, the types of people who report them and the circumstances under which they have been observed, as well as his thorough, impartial style, made him the logical choice for reporting the Hill's account. Of the case Mr. Fuller writes:

I can only add that working with the Hills and Dr. Simon has been a rewarding and educational experience. All three have a passion for accuracy and a profound respect for understatement and documentation. If this comes through in the book, I will have accomplished my objective. (90)

The book, entitled *The Interrupted Journey,* was written five years after an event which had caused the Hills many sleepless nights. Much of the material comprises actual transcripts from tape recordings taken during therapy sessions conducted by Dr. Benjamin Simon in which the Hills were questioned both separately and together, both under hypnosis and fully conscious, for more than six months of grueling investigation. What came out of those sessions is perhaps the most convincing evidence yet that two people were indeed abducted by a group of Starmen, given physical examinations, and told afterward to forget the whole episode.

The anxiety related by the Hills over the incident, both in and out of hypnosis, is apparent throughout the book. Something happened to them which was so terrifying they had vivid nightmares for years afterward. But it was more than a dream, more than an hallucination (even more unlikely since they both related the same experience), and more than some publicity ploy by two people with overactive imaginations.

The story begins on a clear crisp September evening in 1961 as the Hills traveled down route 3, just south of Lancaster, New Hampshire. They were on the return leg of a pleasant vacation spent roving north through Montreal, visiting Niagara Falls and arching back through Portsmouth. The trip was meant to be leisurely. They avoided the crowded major highways, preferring to thread their way through the winding back roads of the White Mountains.

As day turned to evening, the road gradually became more deserted until only an occasional oncoming headlight was encountered. It was about 10:05 p.m. when they left a small cafe in Colebrook, which was about a hundred and seventy miles from their home in Portsmouth. That would put them into Portsmouth by 2:30 or 3:00 a.m. at the latest by their reckoning. Little did they realize how their schedule would soon be drastically altered.

Betty spotted the star-like object first. It caught her attention because it was larger than the surrounding stars and appeared to grow larger and larger by the minute. She finally nudged Barney, who slowed the car to look out the window. At first they thought it was an airplane or perhaps a satellite gone astray. Barney was an airplane watcher, and he figured some people in a Piper Scout were flying through the mountains, though it was rather late to be flying so low. They later estimated that this must have occurred at about 11:00 p.m., though they could not be sure.

They stopped several times along the road to look at the object, which was moving in rather odd patterns. At one of the stops, Betty took out a pair of binoculars and, after some study, decided it definitely wasn't a star or satellite. Near Cannon Mountain they again stopped the car. The object now seemed to have turned from its northerly flight path and was heading toward them. They kept quizzing each other as to possible explanations for the phenomenon. Was it a lost plane, an airliner, a helicopter? Although the object was approaching considerably closer, it remained mysteriously silent. They strained to hear a noise, an engine, a jet rumble, anything which would indicate what they were observing, but could hear nothing.

As the UFO drew nearer, they could begin to make out pencil-thin lights that appeared to rotate around the object, alternating in color from red to amber to green to blue. Barney, who was the skeptic, kept searching for alternative explanations. Betty, who had heard her sister relate a story of a UFO sighting, was more receptive to the new experience:

He was getting irritated at Betty now, or taking out his irritations on her because she was refusing to accept a natural explanation. At one time, several years before, in 1957, Betty's sister and family described seeing clearly an unidentified flying object in Kingston, New Hampshire, where they lived. Betty, who had confidence in her sister's reliability and capacity for observation, believed the story of her sighting. (91)

They got back into the car and continued to drive, more slowly now, looking into the sky with growing apprehension. As they drove, Barney's apprehension turned to fear for the object seemed to have circled around and was following them parallel to their line of travel. As the road curved to the right, the object moved still closer, to about 200 feet or less, in Betty's estimation. Betty was by this time very excited and exclaimed that Barney must stop the car to get a better look at the thing. Reluctantly, he stopped the car. Taking the binoculars, he got out and realized that the UFO had moved to within a height of "two treetops" of them. The silent craft hovered precisely in the night air. It was circular, about three telephone poles in diameter, according to Barney, and had a row of windows which allowed him to make out what appeared to be humanoid forms within.

When his eyes caught sight of the humanoid forms, he became utterly terrified. But despite this, he felt somehow compelled to go around to the other side of the car and out into the adjacent field. It appears he had become mesmerized or hypnotized although at the time he didn't realize it. As he approached to within fifty feet of the craft, it descended to about tree-top level. Betty, still in the car which Barney had inadvertently left parked in the middle of the highway, was screaming for Barney to come back, but he has no recollection of this. He remembers standing out by an old fruit vending stand as the vehicle descended slowly, a few feet at a time, and seeing a pair of what appeared to be wings extending from either side of the thing, each with a red light on the end. From below, something was lowered. He thought it might be a ladder, but again he wasn't sure.

Barney stood frozen in awe. Slowly he raised the binoculars to his eyes. There were *people* or human-like beings standing near the double row of windows which wrapped around the craft! Barney focused on the being in front. He appeared to be the leader. His attire was different from that of the others. He wore a black jacket, a scarf and what appeared to be a captain's hat. The eyes particularly troubled Barney. He likened them to a cat's eyes. They seemed to wrap around the head somewhat as if to give the being several degrees more field of vision. Suddenly he was gripped by the realization of his vulnerability in the open field:

> *For a reason he cannot explain, he was certain he was going to be captured. He tried to pull the glasses away from his eyes, to turn away, but he couldn't. As the focus became sharp, he remembers the eyes of the one crew member who stared down at him. Barney had never seen eyes like that before.* (92)

He ripped the binoculars from his eyes. Like a wild beast about to become prey, he ran screaming back to the car and jammed it into gear, telling Betty he felt they were going to be captured. Betty tried to keep watch as they roared down the highway, but as she peered out the window, she could not see a thing. It was completely black overhead, as if they were in the shadow of the vehicle. At the same time, they heard a strange "electronic beeping" noise coming from what sounded like the trunk area of the car. Associated with the noise they began to feel a "tingling drowsiness" overcoming them, and they become hazy about what happened after that.

The next thing they both recall, which they felt was some time later though neither could guess how long, the beeping sound was heard again, and with the sound they slowly came back to consciousness. The strangest part of their experience would not be uncovered for some time to come, but immediately they knew something was amiss.

They first tried to determine where they were. The last thing they remembered was being a little way south of Indian Head when the beeping started, but the first sign they saw told them they were now some 35 miles *beyond* that, somewhere near Ashland. The Hills recalled the sighting, stopping the car, the field, the strange occupants, but the fuzzy interlude between the beepings would take on a new significance when they reached home and Betty realized that it was *after five o'clock* in the morning! That meant they were at least two hours later than they had originally predicted, and this puzzled them.

There were other things too. They had a strange feeling about themselves. Barney went into the bathroom soon after arriving home to examine his abdomen because the area was bothering him. They discussed the occurrence and decided not to speak to anyone about it for the time being. They had strange impressions or fragmentary recall of a dream-like occurrence as if they had experienced something extraordinary and yet could not summon it to conscious memory. These feelings would haunt them for some time. Only two long years later would they begin to unlock the entire story of the hidden portion of their memories.

In the meantime the Hills began to discuss the experience with a few close friends, Betty being the more speculative, Barney the more skeptical and cynical, but they both became convinced that their experience should be reported. They found that there were apparently numerous UFOs reported in the New Hampshire area about the time of their sighting, and they discovered through friends that they could contact the Pease Air Force Base in Portsmouth to report the incident. Major Paul Henderson was sent to prepare a formal report, which was sent to the Air Force's Project Blue Book in Ohio.

The Hills' anxieties about the event were greatly diminished when they learned of the extent of other sightings being reported by other credible witnesses in the area and through the sensitivity of Major Henderson's inquiry. Barney's primary concern was his being able to justify the sightings in terms of known phenomena. Betty, though sober and cautious, was sparked with a desire to investigate and understand this thing which was thrust upon them.

About ten days after the incident, Betty began to have a series of vivid "dreams" about the UFO with a clarity of detail which she had never before experienced. She had them for five nights in a row and then, as quickly as they had begun, they stopped. These "dreams," it would later be discovered, were memory fragments which temporarily resurfaced from the hazy period of their encounter. At the time of the dreams, Betty refrained from speaking to Barney about them because the subject caused him such anxiety - his only desire was to put the whole thing out of his mind. Fortunately Betty had the foresight to write down what she could recall:

Her dreams were unusual in subject matter and detail. They revealed that she encountered a strange roadblock on a lonely New Hampshire road as a group of men approached the car. The men were dressed alike. As soon as they reached the car, she slipped into unconsciousness. She awoke to find herself and Barney being taken aboard a wholly strange craft. There she was given a complete physical examination by intelligent, humanoid beings. Barney was taken off down a corridor, curving to the contour of the ship, for apparently the same reason. They were assured, in the dream, that no harm would come to them and that they would be released without any conscious memory of the strange happening. (93)

On October 21, 1961, one month after the encounter, a field investigator from the National Investigative Committee for Aerial Phenomena (NICAP), was dispatched to check the Hills' story. Walter Webb, a staff lecturer at the Hayden Planetarium in Boston and scientific advisor for the Washington-based NICAP, occasionally helped check some of the more credible cases in the New England area. His first impression of the Hills and their story was very favorable:

The interview began shortly after noon, and continued with little interruption until eight that evening. "I was so amazed, impressed by both the Hills and their account," Walter Webb later said, "that we skipped lunch and went right through the afternoon and early evening. During that time, I cross-examined them together, separately, together, requestioned them again and again. I tried to make them slip up somewhere, and I couldn't. Theirs was an iron-clad story. They seemed to me to be a sincere, honest couple driving home from vacation, late at night on a lonely road, when suddenly something completely unknown and undefinable descended upon them. Something entirely foreign and alien to their existence." (94)

What transpired on that lonely stretch of highway between Indian Head and Ashland? As investigators and friends learned of the story, and as the Hills recalled it, the focus of attention always came back to the mysterious case of double amnesia. Plagued by agonizing doubts and fears, and keenly interested in coming to terms with their mystery, the Hills agreed to seek professional help on the advice of James MacDonald, a retired

Air Force intelligence officer and personal friend. MacDonald suggested that they might be suffering from a trauma-induced condition not unlike that of "shell shock" or "battle fatigue," and might benefit from medical hypnotic therapy. According to MacDonald, such therapy had produced dramatic results in servicemen suffering temporary amnesia from the severe traumas encountered in war. Subsequent inquiries eventually led the Hills to the man who would unlock the mystery once and for all, Dr. Benjamin Simon.

Dr. Simon, a graduate of Stanford University with an M.D. from Washington University School of Medicine, was a well known Boston psychiatrist and neurologist, having served as chief of neuro-psychiatry and executive officer during World War II at Mason General Hospital, the Army's main psychiatric center at the time. According to Fuller, the necessity of treating up to three thousand patients per month during the war required techniques which could reduce the length of therapy. As a result, one of the many innovative methods which became well established was hypnosis. The use of hypno-therapy was in no small part due to Dr. Simon, who was eventually awarded the Legion of Merit and the Army Commendation Medal for his work at the hospital. After the war Dr. Simon continued using hypnosis for treatment in his private practice.

During the next six months, the Hills would learn of the subtle yet powerful art of hypnosis as they underwent an intensive, detailed investigation of their traumas. All sessions were taped and exhaustively documented - with the Hills together, alone, under hypnosis and fully conscious. To begin the therapy, the Hills were first conditioned to respond to key words spoken by Dr. Simon to facilitate their rapid induction into a hypnotic state upon command. Once this was achieved, subsequent visits could begin straightforwardly and without much repetition.

As the sessions progressed, Betty noted some interesting correlations between the effects of hypnosis in therapy sessions and the trance-like state experienced in conjunction with the UFO incident, particularly on three points: First, she pointed out the loss of awareness of the passage of time while under hypnosis unless instructed to be aware of it. Second, she noted the power of hypnosis to override the body's natural functions. Especially interesting was Dr. Simon's demonstration of the blocking of physical pain or bleeding on suggestion. And third was the power of hypnosis to block off recall of any or all events transpiring during the trance.

The actual therapy sessions began with Barney, who related under hypnosis how he was driving and saw several men on the road standing in front of a bright corona. As Barney spoke, he seemed to actually be reliving the scene, and he often became quite upset, even to the point of screaming out as he did in 1961.

There was real fear and anxiety pent up in Barney as a result of the incident. He recalled with great emotion how some strange men approached his car and opened the door for him. He was instructed, he feels through telepathy, that he would not be harmed, but that he must close his eyes and keep them closed. From the car he was led through a section of forest. He was groggy or dazed. He felt limp though he seemed to be fully conscious. He remembers being carried, being supported at each arm so that his feet dragged on the ground, badly scuffing the tops of his shoes. He was carried up a ramp, and at intervals he cracked his eyelids to glimpse his abductors and the ship into which he was being taken.

Once inside he sensed a temperature change and so cracked his eyelids again to see the inside of the ship. He describes the room into which he was taken. It was "pie-shaped" with the pointed end cut off. Outside the room was a circular corridor, which Betty corroborated in her testimony. Each was led into a separate room for an examination in which scrapings were taken from their skin, fingernails, feet and scalp.

The beings were described as not too different from ordinary men with minor variations. They had mere slits on their faces for noses, and they seemed to have no lip muscles so that their mouths also appeared as slits with straight, short vertical lines at either end. Their eyes were larger and perhaps more elongated, which would have extended their field of vision somewhat. All wore the same type of clothing, described as some type of uniform, though neither could remember what it looked like exactly. The leader was the exception. He was described by Barney as looking like a Nazi with his black scarf around his neck and his jacket and cap. Another feature that stood out was the height difference between the crew and the leader. According to Betty, the leader was about as tall as she while the others were all about five feet tall.

Betty related the same terrified feeling when beings approached her side of the car that night. It appears from her testimony that she was not put so deeply into a trance that she was incapacitated. She felt compelled to do exactly as instructed although she could walk, talk and see all that went on throughout the entire episode.

Betty remembers watching Barney being carried into the spaceship, and she kept telling him, pleading with him, to wake up. At first she was terrified, and then she relates a feeling of being angry at her helplessness. She seems to alternate between the two emotions for several moments but is finally calmed by her abductors. Betty's more-conscious condition allowed her to make the most detailed observations, which were then vividly reproduced under hypnosis. For example, as she is led into her chamber, she relates what she sees: *(under hypnosis)*

... and in one corner, there's a stool, a white - is it white? I don't know if it's white or chrome, but there's a stool, and they put me on it. I sit on the stool. And they - I have a dress, my blue dress on, and they push up the sleeve of my dress, and look at my arm here. They both looked at my arm, and then they turn my arm over and they look at it on here...(indicates a portion of her arm)...and they...they rub it with a machine, I don't know what it is. They bring the machine over and they put it, I don't know what kind of machine, it's something like a microscope, only a microscope with a big lens. And they put - I don't know - they put, I had an idea they were taking pictures of my skin. (95)

Betty related these things while under hypnosis and therefore had a tendency to be completely accurate, even to the point of being somewhat repetitive. The examination was very thorough. The aliens inspect her eyes, ears, nose and throat, taking samples and placing them into separate containers. They take samples of hair, skin and fingernail. The entire procedure was described in some detail and appears quite comprehensive by any measure. They then have her remove her dress. They examine her hands and feet in more detail, they use an electronic device to check the nervous system (Betty is told by the Starmen what the device is used for), and they even give her a pregnancy test!

...They roll me over on my back, and the examiner has a long needle in his hand. And I see the needle. And it's bigger than any needle that I've ever seen. And I ask him what he's going to do with it...(She begins to get upset again). It won't hurt me. And I ask him what, and he said he just wants to put it in my navel, it's just a simple test. (More rapid sobbing) And I tell him no, it will hurt, don't do it, don't do it. And I'm crying and I'm telling him, "It's hurting, it's hurting, take it out, take it out!" And the leader comes over and puts his hand, rubs his hand in front of my eyes, and he says it will be alright, and I

won't feel it. (She becomes calmer) And all the pain goes away. The pain goes away, but I'm still sore from where they put that needle... (96)

When Betty had completed her series of tests, she asked to go back to the car, but the leader indicated that Barney had not yet finished with his testing. Betty was grateful to the leader for stopping the pain, and she began to ask questions about her abductors. She asked, for example, where they came from, whereupon she was shown a stellar map with lines, dots and circles, but she could not read the map. She had no knowledge of astronomy, and so the map was put away. Later under hypnosis, Betty redrew the map from memory. At last she felt she could produce some definite proof that the incident really had occurred. The map was subsequently analyzed by astronomers, but at that time there was no known star pattern which fit the map Betty had drawn.

Some years later, however, an amateur astronomer named Marjorie Fish was successful in orienting the map to discover its implications. Today the map provides an interesting corroboration to the Starman Intervention Theory and will be addressed in a subsequent section.

Meanwhile Betty especially wanted some artifact, some bit of evidence for proof because nobody would believe her when she told them about the encounter. At first the leader had allowed her to take a book written in their alien language, but later she was forced to leave it behind when other members of the crew complained that the experience was supposed to be wiped from conscious memory. She was told it would do her no good anyway since all materials and script could be duplicated on earth.

At one point, even in the serious overtones of this whole event, there was a moment of humor:

All of a sudden, there's this noise in the hall. Some of the other men came in and with them is the examiner. They are quite excited, so I ask the leader what's the matter with them. Did something happen to Barney? It has something to do with Barney. The examiner has me open my mouth, and he starts checking my teeth. And they are tugging at them. I ask what they are trying to do...they are tugging, pulling at them. They are very excited. (She laughs) The examiner said they could not figure it out - Barney's teeth came out, and mine didn't. I was really laughing, and said Barney had dentures, and I didn't, and that's why his teeth came out. (97)

There ensued a conversation about what dentures were, why some people needed them and some did not. The leader and Betty spoke of many things, some of which the beings had difficulty grasping, such as the concepts of age, the passage of time, colors, etc., at least by Betty's explanations.

Finally the Hills are led out of the vehicle, back through the woods and to their car. They watched the vehicle leave as silently and precisely as it had arrived. And they would forget, just as they were told they would. As they drove away and entered the highway again, Betty remembers asking Barney if he believed in flying saucers now, and Barney replied, "Don't be ridiculous." Next they remember hearing the second set of beeps, and they began looking for a sign to find out where they were, for they had already forgotten what had just taken place.

* * *

It is significant that the contact of individuals or small groups is the preferred method of interaction chosen by the alien beings. Their missions appear to involve monitoring and sample gathering rather than overt, formal communications with official governments, as is demonstrated in the cases thus far. While there is no universally accepted answer as to why this method is used, it is nevertheless tempting to speculate that other intelligent races visiting the earth would look at our level of sophistication and recognize that we are not yet mature enough as a civilization to make communications profitable on an official basis.

As will be shown, however, Starmen did once deal with an entire race of people, as is recorded in the Bible. And it is quite possible that we are now on the threshold of commencing such communications again. Our ability to move permanently out into space is near, and if such alien life exists and plies the heavens, then we must prepare to enter their domain on their terms.

Carl Sagan, in his highly acclaimed television series *Cosmos*, spoke of the tremendous possibilities for the existence of intelligent life on other planets and of the possible galactic order which might await our initiation into space. He even once openly speculated that we may have been visited many times in the past by these beings although he has since tempered his past statements with the revised assertion that we still do not have any concrete proof that such beings have actually visited the Earth. In an address to the American Astronomical Society in 1966, Sagan stated:

...the earth may have been visited by various galactic civilizations many times during geologic time. It is not out of the question that artifacts of their visits still exist or even that some kind of base is maintained within the solar system to provide continuity for successive expeditions. (98)

Sagan has also said that beings from another world would not, in all probability, evolve to look anything like us for evolution has many different pathways. But there are those who disagree:

Bieri, (1964) argued that extraterrestrial intelligence probably will resemble homo sapiens because evolutionary pathways are strictly limited and air breathing, land roaming humanoids offer the optimum adaptive solution to terrestrial environments. He postulated that life-supporting planets will evolve bilaterally with symmetrical animals with anterior brains and closely associated sense organs, paired appendages, and hands for tool making. (99)

Another aspect...concerns the idea that [alien] visitors and their equipment...would have to be and look different from us and from whatever we use. The tendency to attribute a priori to an extraterrestrial civilization fanciful, mysterious, or unfamiliar shapes and capabilities often overlooks the proposition that a resemblance is in fact much more probable than a substantial difference. (100)

Concerning the Hill case, Sagan has stood on the *Cosmos* set and with a wave of his hand said that the incident was interesting, but that there wasn't "a shred of evidence that this event occurred" and that all we have is two people's word. Again herein lies the paradox of Sagan's philosophy for he states that he believes in the possibility of intelligent extraterrestrial life, and yet he is skeptical that aliens have visited this planet and rejects that they would look like humans.

But there actually is some interesting corroborative evidence to support the Hill abduction story. One of the most important pieces of evidence to come out of this case involves the star map which Betty Hill drew while under hypnosis and which she says was shown to her by the leader of the Starmen during her abduction. It supposedly shows a group of stars connected by various arching lines representing travel routes.

As stated earlier, the map provoked no real stir in the scientific community when released. Indeed, Carl Sagan, on his

Cosmos television series, said the lines drawn to connect the various stars illustrated by Betty Hill could have been drawn many different ways to come up with completely different star systems and thus was useless as a reference. Whether Sagan knew of the theory worked out by amateur astronomer Marjorie Fish regarding the Hill map is unknown, but Fish's results certainly bear out a different conclusion, which effectively limited the number of possible star combinations to *one.*

Marjorie Fish's Stellar Map

Marjorie E. Fish, a thirty-four year old elementary school teacher and amateur astronomer from Oak Harbor, Ohio, in 1966 set out to discover if the map produced by Betty Hill could possibly represent an actual star pattern somewhere in space. Marjorie believed the Hills' story could be true. She also felt that if the account was true, the map Betty drew must represent an existing star pattern, and so she embarked on a project that was to take her six years of study and involve over 20 models of our stellar neighborhood. The task was immense. Her star models consisted of up to 259 beads suspended on strings, each colored and sized to represent its actual counterpart in space. She fully expected to find many star patterns which fit the Hill map, but to her amazement only *one* model emerged which fit both our stellar neighborhood and Betty Hill's map.

There were four natural assumptions used by Fish in her analysis, the first being that the Starmen's home star must be represented by one of the large circles in the foreground of Betty's map; second, that the orientation of the map was from the vantage point of the Starmen's home star; third, that our sun was included on the map, with one of the several lines drawn to it, and fourth, that the map therefore depicted our stellar neighborhood. With these constraints in mind, we shall return for a moment to our layman's astronomy to understand how Ms. Fish discovered the valid model.

Recall in an earlier chapter that we discussed how a star comes into being when a cloud of space dust gains sufficient mass (density and weight) to exert the necessary pressure upon itself to ignite. Once ignited, it will burn at a constant rate for a great period of time (called main-sequence), and then eventually burn out or even explode, depending upon its mass. Stars form in all sizes, each giving off a unique combination of color and intensity of light, and it is in this way that astronomers classify them, according to their "spectral class."

There are seven primary spectral classes, O, B, A, F, G, K, and M, each in order of their decreasing surface temperature. O stars are large, blue and hot. G types are yellow, medium sized and medium temperature, and M stars are small, red and cool. Our sun is a G spectral class star. The designations are further broken down into ten increments, 0 through 9, in order of decreasing temperature for more precision. Therefore a star of spectral class G5 would be hotter than an M1 or G9.

Fish made the assumption early on that extraterrestrial space travellers would choose to visit primarily stars similar to their own, those offering the greatest probability of encountering life, much the way we would probably search out the life-bearing stars could we travel the great distances of space. She therefore chose those stars defined by expert astronomers such as Sagan, Dole, Oliver and Huang as having the most probability of having planets capable of generating intelligent life.

The presence of planets around a star is indicated by the star's rotational speed. Planets affect a star's rotation by draining some of its angular momentum, thereby slowing it down.

Also, in order for intelligent life to develop on the planets surrounding a star, it must, among other things, have been burning for the right amount of time and it must be stable; that is, it must not send out into space huge solar flares, which would destroy the life on its planets. Therefore the stars that Fish studied all had to be, by her own criteria, non-fluctuating main-sequence stars between F8 and K1.

With these facts in mind, we discover some amazing findings of the Fish project. First, in her entire six-year search, she found but *one* three-dimensional duplicate of the Hill star map. No other pattern even remotely resembles it among 200-odd stars in the F8 to K1 spectral range which reside within 72 light years of the sun. Second, the stars that make up the pattern in the Fish model are all within the aforementioned criteria for supporting intelligent life. Third, and most interesting, if the map was somehow faked (and Betty claims no knowledge of astronomy), it could not have been accurate as drawn in 1964 because three of the stars would have been misplaced or missing. Prior to the 1969 edition of Gliese's *Catalog of Nearby Stars,* one of the stars, Gliese 86.1, was not listed in any existing star catalog, and two others, Gliese 95 and

Kappa Fornacis had improper parallaxes and would not have formed the background triangle of stars shown on the Hill map. (101)

In astronomic terms, the stars contained in the ultimate Fish model are described as follows:

The twelve stars connected by lines have the following ranges as a group; spectral classes F6 V to K1 V (Sun G2 V), surface temperatures about 4800 to 6500 degrees Kelvin (Sun 5800 degrees), masses about 0.7 to 1.2 solar masses (Sun 1.0), radii about 0.7 to 1.2 solar radii (Sun 1.0), absolute visual magnitude plus 3.7 to plus 7.0 (Sun minus 26.7), distance from earth 11.8 to 52.6 light years, and main sequence residence times about 7 to 30 or 40 billion years (Sun about 13 billion). (102)

Therefore, the possible origin of the Starmen in the Hill case: Zeta Reticuli, which is thought to be a star very similar to our own. Detailed analysis of the Fish model by astronomer Walter N. Webb and others thus strongly challenges Sagan's assertion that Betty Hill's star map could represent any number of star combinations. Carefully selected criteria and meticulous scrutiny of the available information suggest just the opposite; that the sun is one of the stars shown on the Hill map, that we are being monitored by Zeta Reticulans and others on their stellar expeditions, and that the abduction of Barney and Betty Hill represents an actual occurrence.

Endnotes

74. Jacques Vallee, *Passport to Magonia,* Henery Regnery Co., Chicago, 1969, p. 12.

75. *Ibid.,* p. 10.

76. *Ibid.,* p. 15.

77. *Ibid.,* p. 13.

78. Travis Walton, *The Walton Experience,* Berkley Publishing, New York, 1978, p. 8.

79. *Ibid.,* p. 70.

80. *Ibid.,* p. 76.

81. *Ibid.,* p. 109.

82. Bill Barry, *Ultimate Encounter,* Pocket Books, New York, 1978, p. 113.

83. *Ibid.,* p. 117.

84. Raymond E. Fowler, *The Andreasson Affair,* Bantam Books, New York, 1979, p. 5.

85. *Ibid.,* p. 4.

86. *Ibid.,* p. 13.

87. *Ibid.,* p. 21.

88. *Ibid.,* p. 112.

89. *Ibid.,* p. 119.

90. John G. Fuller, *The Interrupted Journey,* Berkley Medallion Books, New York, 1966, p. 15.

91. *Ibid.,* p. 27.

92. *Ibid.,* p. 32.

93. *Ibid.,* p. 49.

94. *Ibid.,* p. 52.

95. *Ibid.,* p. 192.

96. *Ibid.,* pp. 195-6.

97. *Ibid.,* p. 209.

98. Blum, *op. cit.,* p. 203.

99. Lorenzen, *op. cit.,* p. 109.

100. Josef F. Blumrich, *The Spaceships of Ezekiel,* Bantam Books, New York, 1974, p. 6.

101. Lorenzen, *op. cit.,* p. 108.

102. *Ibid.*

Genesis 28:12

Chapter VII
THE BIBLE

...The way God has been thought of for thousands of years is no longer convincing; if anything is dead, it can only be the traditional thought *of God.*

- Hannah Arendt, 1970

In our examination of The Order, it has been proposed that both religion and science emerged from the dim reaches of man's quest to understand, to explain and to influence the world around him. It was proposed that religion is a spontaneous phenomenon, as evidenced by the historical and continued emergence and evolution of new cults and religions today. In addition, it was pointed out that the basic religion of the Hebrews was quite different from other religions, including Christianity, in that the god they worshiped was a "living" god - a deity who was, in fact, a Starman.

Visits from Starmen probably have occurred at times and locations on the Earth other than the series of incidents recorded in the Bible. Traditions and legends from the Indus Valley region, China, Australia, Egypt, Persia and the Americas all contain intriguing allusions to "gods" from the sky who came down to Earth and imparted wisdom, culture and civilization on mankind. The Hebrews themselves seem to have originated from the region around what is now known as ancient Sumeria, the supposed "cradle of civilization" which, in turn, has its own complex theogony, some of which appears to have been drawn upon by the Hebrews. Therefore the story related in the Bible is already merely the essence of one period of Starman visitation, the earliest contacts occurring, according to some authorities, before the rise of man.

In laying the foundation for the Starman Intervention Theory, it has been shown how evolution produced many cultures on other worlds, some of which may have visited Earth, and sporadically continue to visit in what we call UFOs. Initially it was shown how the Bible as now interpreted gives a distorted picture of these events so that Christianity, as practiced today, is based on some mistaken precepts about history and God. Finally, we shall now turn to the story of the Bible as it *should* be interpreted.

It is proposed that, with the exception of the earliest portions of the Old Testament story, (which appear to be drawn from still more ancient sources), the Bible records an actual series of extraterrestrial interventions into Hebrew culture. The resulting "religion" which arose from those interactions has locked in verse knowledge which we can now begin to understand for what it really was and thus put our perceptions of the order of things in their proper perspective.

Overview

At the outset of an examination of the Bible under the Starman Intervention Theory, we must state three basic premises. The first premise about the Bible is that it contains a series of accounts which concern a single race of people - the Hebrews - and their special relationship to their god. Moreover, the god of the Bible does not concern himself with any other race. The Hebrews were his chosen people, and they alone are the focus of his attention.

The second premise about the Bible which must be underscored is the fact that it is a collection of 66 books written by many different authors over an expanse of hundreds, even thousands of years. As stated earlier, the Bible has evolved through a long and often stormy history. Rutherford H. Platt explains:

The compilation of the Bible was not an act of any definite occurrence. It was a matter complicated and abstruse. It was an evolution at the hands of Churchmen of various beliefs and purposes. In the formation of early Church doctrines there was dissension, personal jealousy, intolerance, persecution, bigotry. (103)

From Genesis to the Gospels, virtually every book of the Bible has evolved from diverse sources under various other names. Politics have played a substantial role in shaping the content and format of the text. Portions of books, and in some cases entire books, have been omitted from the canon. The non-canonical writings are called the Apocrypha and the Psuedepigrapha. These form a large body of material which modern scholars consider an intimate part of the history of the formation of the Bible and help shed light on the theology which shaped the scriptures. We shall not exclude them from consideration.

It should also be noted that some narratives, such as the story of Noah and the Ark, are derived from several different ancient versions which were then compiled at a relatively recent date. This sometimes resulted in conspicuous discrepancies. In the case of the Noah story, for example, the editor seems to have been uncertain of the correct version of the story. Not knowing which parts to include and which to leave out, he simply overlapped them, one right after the other, creating a conflicting report. In one passage of Genesis it reads that it rained for 40 days and 40 nights and then the waters of the flood abated (Gen. 8:6-12), whereas another passage reads that the flood waters abated after 150 days (Gen. 7:24).

The two different versions of the flood story are one example of how conflicting source documents have been compiled to form the Bible. Each of the source documents has been classified by scholars according to its origins and reflects the distinct geopolitical, social and religious conditions which prevailed at the time it was authored.

It is beyond the scope of this book to present more than the barest of details, but it may be helpful to list the major documents to demonstrate the diversity of sources. They include: The *J Document,* written by a Judean who lived in Jerusalem circa 950-850 B.C.; the *S Document,* thought to orginate from Edom during the first half of the tenth century B.C.; the *E Document,* composed in northern Israel about 750 B.C. by a priest; the *D Document,* found by Hilkiah, the High Priest, in the collection box at the entrance to the temple of Jerusalem in 622 B.C; and the *P Document,* or Priestly Code, written by a priest in Jerusalem c. 450 B.C.

In addition there are other so-called "codes" and "fragments" such as the *Covenant Code,* which dates from about 550 B.C. (104) From these the Bible evolved. They are by no means to be found in any order in the Bible. As stated previously, they have been found overlapped, added to and edited to produce the Bible we recognize today.

The third premise about the Bible is that its theme is not, as many theologians assert, the story of mankind leading up to the coming of Christ:

The central theme of the Bible is Christ. It is this manifestation of Jesus Christ, his person as "God manifest in flesh," his sacrificial death and his resurrection, which constitutes the Gospel. Unto this all preceding Scripture leads, from this all following scripture proceeds. (105)

On the contrary, it is inescapably evident that the *central* theme of the Bible is the establishment of the Hebrews in their own land (Canaan) that they might become a race of priests unto the rest of mankind. The Hebrews' obligation to God was cemented with a covenant, an oath of loyalty. This involved a promise by God to establish the Hebrews in a land of their own where they would become exalted among all mankind and a promise by the Hebrews to obey God and all his ordinances through faithful service, thus becoming the instruments through which God would deal with all mankind. However, when it became clear to God, after centuries of Hebrew abuse of this sacred covenant oath, that the Hebrews could not be "tamed," Jesus was sent to *break* the covenant, thus ending the special relationship between God and his people.

There are roughly three main divisions to the activities of the Bible. They are: First, the Old Order, the ancient times when the Starmen walked freely among men as "gods." It was the period when the Hebrews were actually born as a race and were kept separate from other races in the "garden" of Eden. At this time their only service was to God. When God's trust was broken by these early Hebrews, they were exiled from the garden, and a terrible degenerative force was set in motion, culminating in the destruction of their world by a deluge.

Second, the Intermediate Order, its beginning marked by God's declaration of a mission for the Hebrews and his commissioning of Abram as the first proxy leader. It was during this time that the Hebrews were led on the conquest of Canaan through such patriarchs as Moses, Joshua and David. It was a time when awesome behind-the-scenes forces were at work ensuring the success of the Hebrews in their quest for a land of their own.

Third, there was the end of The Order, the breaking of the covenant, the departing of the Starmen, leaving a destitute culture, splintered and impotent. Thereafter a whole new and quite different order was spontaneously generated in the wake of the political debris of the rejected Hebrews. From then on the precepts taught by Christ became the "bread" of the masses, and the resulting momentum propelled a new cult, Christianity, to the forefront of the world's greatest religions, all through a spectacular series of political, social and cultural vacuums. The new religion was thus born through an intriguing series of Hebrew defaults on their ancient covenant with the Starmen.

There is one final consideration to be made regarding the level of credibility to be placed upon the stories of the Bible. In as much as possible, a great effort has been made to get to the root of the various stories, to classify them according to expert opinion where possible into historical or mythical categories. From these was drawn a third category, that of phenomena which could have been misidentified because of an alien presence in accordance with our assumption of an extra-terrestrial intervention. For this "misidentified phenomena" category, it is assumed that what was being related actually occurred, and an attempt is made to understand the event in terms of modern-day knowledge. In essence, it is the taking of the stories literally for what they *really* are.

Meanings of Words

Not only has the Bible evolved through a long series of physical changes, but the ideas behind the words have undergone a metamorphosis as well. There are primarily two reasons for this. First, there was the evolution of the language itself, accompanied by mechanical transcribing errors and corruptions of word meanings as described in the preface to World Publishing's *R.S.V. Bible.* (1962):

Sometimes the present translation will be found to render a Hebrew word in a sense quite different from that of the traditional interpretation... If in the judgment of the Committee the meaning of a passage is quite uncertain or obscure, either because of corruption of the text or because of the inadequacy of our present knowledge of the language, the fact is indicated by a note. (106)

Such a word is "Jehovah," which is medieval corruption of the original Hebrew consonants YHWH. The consonants were symbolic for the name of the Lord but were not meant to be pronounced because of the sacredness of the name. Behind that mystic tradition was the idea that to know and invoke a name was to give an individual a certain power over the person named, and thus the true pronunciation was never revealed.

As the text was handed down through the aeons, the language of earlier generations became obsolete to those of later ones. To continue quoting from the *R.S.V. Bible* preface:

The greatest problem, however, is presented by the English words which are still in constant use but now convey a different meaning from that which they had in 1611 and in the King James Version... The King James Version uses the word "let" in the sense of "hinder," "prevent" to mean "precede," "allow" in the sense of "approve," "communicate" for "share," "conversation" for "conduct," "comprehend" for "overcome," "ghost" for "spirit," "wealth" for "well-being," "allege" for "prove," "demand" for "ask," "take no thought" for "be not anxious," "purchase a good degree" for "gain a good standing," etc. (107)

The editors go on to say that there are over three hundred such words in the King James version alone. Now, with the proliferation of new Bible versions such as the *Good News Bible,* the *Living Bible,* and the National Council of Church's new addition, the *Neuter Bible,* we are constantly reminded of the dynamic nature of language and cultural influence on the Scriptures.

But there is yet another type of change which we must be aware of. In addition to the evolutionary changes in the common meanings of words, two additional sources of confusion have arisen. One is broadened interpretation of some words resulting from man's increasing knowledge, and the other is, according to our thesis, the attempt by an ancient culture to describe in familiar words a technology beyond their understanding.

Recall the earlier assertion that it was not the entire world which God was concerned with; it was the Hebrews only. Consider also that the "world" to the ancient Hebrews was their Middle Eastern location from Palestine to eastern Mesopotamia. They did not, in general, know of such far off lands in China or the Americas.

To the Eastern people of the earliest times the world was very small, including little except Mesopotamia, Canaan, Arabia, and parts of Egypt. During Old Testament times it remained much the same, extending a little to the East, as the nations there grew to power and other parts of Asia and Africa became somewhat known to the adventurous. Few knew more than rumors of Italy and even Greece. In the times of Christ the world really meant the Roman Empire, with parts of Asia and Africa which were more or less under its sway. All outside of this was vague. (108)

The meaning attached to such a word, therefore, depends on a degree of knowledge. Similarly, there are many words used in the Bible which could be termed "words of comparison." These words represent the attempts of a primitive culture to describe elements of a scientifically superior one and gave us such key terms as "visions," "holy," "glory," "spirit," and "cherubim." This is an important idea for words used to convey ideas depend heavily on the context in which they are used. Obviously differences in technology would cause concepts derived from the words of one culture to convey different meanings to people of different levels of sophistication.

Fortunately such words, if preserved in their comparison forms, would only remain "coded" for as long as that culture was technologically inferior to the superior one. It is our present technological level, therefore, which allows us to examine the text of the Bible and see, through our advanced perspective, that what the ancients *really* saw was a foreign technology intervening into their culture and not a "spirit" god. We can now understand their descriptions of things which they themselves did not fully understand and how they reacted to that presence; they built their religion around it!

All of the many words which could fit into the category of "comparison words" cannot be detailed adequately here but will be covered as necessary throughout the text in order that they may be appreciated in context.

Endnotes

103. Editors, Rutherford H. Platt, Jr., *The Lost Books of the Bible and the Forgotten Books of Eden,* William Collins Pub., Inc., 1926, 1927, from the Preface.

104. Scofield, Rev. C. I., D. D., *The Holy Bible,* Oxford University Press, New York, 1945, p. vi.

105. *Ibid.*

106. Editors, *The Holy Bible, R.S.V.,* World Publishing Co., 1952, p. v.

107. *Ibid.,* p. vii.

108. Alexander A. M. Cruden, *Cruden's Complete Concordance,* Zondervan Publishing House, Grand Rapids, MI, 1968, p. 762.

Ezekiel 43:1-7

Chapter VIII
THE OLD ORDER

O Lord my God, thou art very great!
...who coverest thyself with light as with a garment
...who makes the clouds thy chariot,
who ridest on the wings of the wind
— Psalms 104:1-3

Mesopotamia

Mesopotamia is considered the cradle of civilization, birthplace of the great Sumerian culture. The word *Mesopotamia* is Greek, meaning "the land between the rivers." These two life-giving rivers, the mighty Tigris and Euphrates, flow through what is now a part of Iraq and Syria to a fertile delta that empties into the Persian Gulf. Since time immemorial Mesopotamia has been an oasis for primitive bands of wandering nomads, providing a respite from the often harsh region surrounding it. There, in the heart of what is called the "fertile crescent," the low grassy land provided excellent grazing for sheep and goats, easy access to water for the irrigation of fields and transport, and a climate hospitable to both crops and humans. It was, therefore, a natural crossroads for man, a place where he found good reason to congregate and advantages to settling down. So powerful was its attraction that the nomads were eventually enticed to give up their wandering ways and settle there permanently sometime between 11,000 and 10,000 years ago.

But all of mankind was not restricted to this Middle-eastern valley only for humans had long since spread far and wide across the globe. Moreover, they were already well on their way to diversifying into completely unique variants, or races, within their ever expanding domain. The cause of these different races, according to Leakey, was:

Evolutionary spurts influenced by local conditions produced local variants, this much we know. For instance, the robust australopithecines that lived along the side of Lake Turkana were substantially stockier individuals than those that lived in southern Africa. Similar kinds of variation in build and size developed among the Homo erectus populations spread throughout Europe, Asia and Africa a million and more years ago. You only have to look around the world today to see that, although humans are indisputably members of the same species, there are distinct geographical characteristics. (109)

If, as anthropologists argue, man grew intellectually and socially to the point of spontaneous culture emergence and civilization, then it is interesting that the first civilization arose in the same location as described in the biblical narrative of Adam and Eve and their Garden of Eden.

Many critiques of the evolutionary theory of man involve the rapidity with which man progressed from his primitive tool-making industries to the elaborate and sophisticated civilizations. Some have estimated that it took some 2,000,000 years for the earliest tool makers to refine their tool industry to the point of the deftly crafted flaked stones and flints of the late stone age, and yet there are perhaps less than 50,000 years separating Neanderthal Man from the landing of the astronauts on the moon! (110)

Rather suddenly, between 4,000 and 3,000 years ago, a magnificent civilization arose from the land between the two rivers. "Mesopotamians were the first people on earth to live in cities, study the stars, use the arch and wheeled vehicles, write epic poetry and compile a legal code." (111) Is it a coincidence that the most advanced peoples of the world by far developed in the very same place from which the Hebrew God of the Bible placed his garden when he created Adam and Eve? Clearly theologians and scientists have another point of common ground since this is where the biblical scholars locate the Garden of Eden as they deduce from the scriptures:

And the Lord planted a garden in Eden, in the east...A river flowed out of the garden, and there it divided and became four rivers...The name of the first is Pishon...The name of the second is Gihon;...And the name of the third river is Tigris, which flows east of Assyria. And the fourth is Euphrates. (Gen. 2:8, 10-14)

Although scholars have disagreed on the locations of the first two rivers, there can be no doubt of the locations of the Tigris and Euphrates. The report that the garden was planted in the east is matter-of-fact and to the point, as if the garden really existed. Was the story of Adam and Eve a real occurrence?

Obviously, the first part of Genesis is of remote antiquity. What is told of that old order by the authors of Genesis narratives was a distilling down of what was already ancient history even to them. For the thesis of this book, the *Old Order* corresponds roughly to the period from the arrival of the Starmen through the time of the great flood and perhaps a little afterward. This period deals with the creation of the Hebrews and their face-to-face relationship with God and his angels, (who at this time are portrayed as walking among the gardeners in physical form) and who serve God in blissful ignorance until they were led astray by the satan element, for which they suffered expulsion from the garden and the toil of the "outside" world. This is the Eden metaphor because although it describes this old order in a rather simplistic way, there is evidence that there really existed a place called Eden.

The Eden metaphor commences with God's creation of man and his placement in the garden, where he was innocent and pure in his ignorant state. Very little is revealed about Eden in the Bible beyond this. The story of Eden was significantly altered by early churchmen into a description of man's fall from a state of divine grace through an act of disobedience to God's will. The burden of this "sin" was then placed squarely on man's shoulders when, in fact, the man was but a pawn in a battle among the angels. From certain apocryphal books, we are able to draw additional details about the Eden metaphor not found in the Bible and which help reconstruct what might have really happened so long ago in that dim past. A key element in the Eden metaphor concerns the symbolic trees for which the gardners had special instructions:

And out of the ground the Lord God made to grow every tree that is pleasant to the sight and good for food, the tree of life also in the midst of the garden, and the tree of knowledge of good and evil. (Gen. 2:9)

The "tree of life," like the familiar "family tree," represented the gardeners' privilege, as the biologically created offspring of the Starmen, to enjoy the Starmen's long lifespans. It was literally a "family tree - of life!" The Starmen, as an advanced civilization, had probably long before discovered the secret behind the mystery of aging. Through manipulations of their biological and chemical structures, they were able to extend their lifespans to upwards of one thousand years, (as evidenced by the near millennium lifespans of the early patriarchs). Therefore as long as the gardeners served under the Lord, they could expect to have lifespans which, to us, seem like forever. Consider for a moment our current understanding of aging.

In the Old Testament times the mean life expectancy was under 40 years. (112) From the tolls taken by diseases, wars, famine and poor nutrition, one was considered lucky to survive beyond puberty. Today we have conquered many of the diseases which caused so many of the deaths in earlier times, and we have learned that exercise, proper diet and environment can help us live to seventy or more years of age. But what of the incredible ages reported in the Old Testament? Are we to take them literally? We read reports that Adam lived for 930 years, Seth for 912 years, and about the great granddad Methuselah, who lived to be 969 years old! In all, the reported ages of the pre-flood patriarchs are as follows:

1.	Adam	930 years	6.	Jared	962 years
2.	Seth	912	7.	Enoch - taken into heaven	
3.	Enosh	905	8.	Methuselah	969
4.	Kenan	910	9.	Lamech	777
5.	Mahalel	895	10.	Noah	950

It is interesting to note, too, that after the flood the lifespans of the patriarchs gradually decline to more familiar levels:

1.	Shem	600 years	9.	Terah	205 years
2.	Arpachshad	438	10.	Abraham	175
3.	Shelah	433	11.	Isaac	180
4.	Eber	464	12.	Jacob	147
5.	Peleg	239	13.	Joseph	110
6.	Reu	239	14.	Moses	120
7.	Serug	230	15.	Joshua	110
8.	Nahor	148			

(Genesis 11:10-32)

What does this gradual decrease indicate? According to the Bible, this genealogy is a father-to-son tracing from Adam to Joseph! Are such lifespans possible?

Today we are talking of pushing the age limits of man from 74+ for males and 78+ for females to perhaps as high as 100 in the near future. What remains to be answered by science today is the mechanism which triggers the aging process. Current theories of aging are beginning to focus on the idea that all organisms live according to a kind of biological clock which genetically programs the rate at which their cells age. As time passes, the individual cells which make up the organism begin to lose their vigor and their ability to resist invasion by diseases.

Some scientists believe that aging is caused by special hormones released in the body according to the biological clock. Others feel that there is an inherent factor in the genetic material of each cell which simply turns off maintenance and repair cycles, allowing the cells to wear out and repair themselves too slowly to sustain life. Perhaps the answer is a combination of the two.

A University of Nebraska biochemist, Denham Harman, believes that *"...aging results from wear and tear inflicted by 'free radicals,' broken-off cellular debris that clusters on cells, clogging or rusting them out of action."* (113) His experiments showed that diet agents, such as the chemical antitoxidant compound vitamin E, would cause lifespans of mice to increase by an average 49 per cent. Other studies have shown that administering certain drugs to restrict the effects of various glands have had significant results. In one experiment by W.D. Denkla, the pituitary gland was surgically removed from groups of test rats:

Close monitoring showed that a variety of rats' body functions had returned to "juvenile competence." Their coats were again thick and glossy, their movements were quick and energetic. The rats showed no signs of age or debility until almost the point of death. (114)

The efforts of scientists today, though primitive, will in the near future, perhaps in our children's lifetimes, yield the secrets of the process of aging. Not only will we live significantly longer, but we will be able to stretch out the early and middle years, virtually eliminating the old age segment and allowing people to be fully productive longer. Looking at our future in the field, we may see that the lifespans of the pre-flood patriarchs can be explained in terms of the superior technology of the Starmen. If the Starmen had conquered the age barrier, then their created man would be blessed with that trait also. This was the family "Tree of Life."

Of the other tree, the "Tree of Knowledge of Good and Evil," the gardeners were strictly forbidden to "eat." As the Tree of Life was a metaphor for the Hebrews' special family tree, similarly the Tree of Knowledge represented a special source of knowledge forbidden to the gardeners. Was the fruit of this tree literally eaten? Probably not for, as some argue, the word "eat" in the sense that Eve ate the "fruit" of the tree means "to acquire" or "to learn":

As an aid to understanding the Hebrew of the Bible, there are special dictionaries in which each word used in context is followed by an indication of all the other passages where it occurs. The Hebrew word usually translated as "to eat" often occurs in context where it can only have that meaning, but it also occurs in context where its meaning is obviously different. In Ezekiel, for example, a voice orders Ezekiel to "eat" a scroll, then go teach what he has learned by "eating." [Ezekiel 3:1] (115)

As examples of similar usages today, we sometimes use the term "to digest" material we have read, to ponder it, and also to find ideas "palatable" or not. Eve, therefore, was tempted by someone, depicted as a cunning serpent, to learn of this forbidden knowledge. In order to more fully understand the concept of the temptation of Eve, we must jump ahead for a moment. The non-canonical Book of Enoch, thought to have been written between 130 and 100 B.C., is a rich source of information on this matter. Perhaps one of the reasons for its exclusion from the Bible was that it describes the corruption of man by a band of angels. Only fragments of the Enochian text remain intact in the Bible. Genesis 6:1-4 is one of those fragments:

When men began to multiply on the face of the ground, and daughters were born to them, the sons of God saw that the daughters of men were fair; and they took to wife such of them as they chose...The Nephilim were on the earth in those days, and also afterward, when the sons of God came into the daughters of men and bore children to them. These were the mighty men that were of old, the men of renown.

The traditional translation of the Hebrew word *Nephilim* is "giants" but, according to several other sources, it could also be translated as "fallen ones" or "those who were cast down." (116) To corroborate this idea, we find in Revelation 12:7-9 that it was *Satan* who was cast to earth:

Now war arose in heaven, Michael and his angels fighting against the dragon; and the dragon and his angels fought, but they were defeated and there was no longer any place for them in heaven. And the great dragon was thrown down, that ancient serpent who is called Devil and Satan, the deceiver of the whole world - he was thrown down to earth, and his angels thrown down with him.

Now, when we read the portion of the Book of Enoch which corresponds to the Bible version, we may gain an understanding of what really happened:

And it came to pass when the children of men had multiplied that in those days were born unto them beautiful and comely daughters. And the angels, the children of heaven, saw and lusted after them, and said to one another; "Come, let us choose wives from among the children of men and beget us children." ...And they were in all two hundred; who descended in the days of Jared on the summit of Mount Hermon. (Enoch VII:1,6)

These angels who were eventually cast down to earth for their corruption of mankind were called "Watchers." They originally had the responsibilty to watch over and monitor the Earth. But when some of these Watchers decided to come down and take wives from among Earthmen, they were destroying a greater plan by the Lord and his good angels. According to Enoch, it was a Watcher who led Eve to "sin":

And the third was named Gadreel; he it is who showed the children of men all the blows of death, and he led astray Eve...(Enoch LXIX:6)

We may now return to the concept of the temptation of Eve and see that it was the Watcher Gadreel who is portrayed in the Eden metaphor as Satan, the deceiver, who penetrated the Garden unbeknown to the arch angels to tempt Eve by saying, *"Did God say, 'You shall not eat of any tree of the garden?'"* (Gen. 3:1)

This war in heaven between the Lord's angels and Satan's is thus portrayed in Genesis. They were physical beings capable of intercourse with humans and engaged in a physical war. The temptation of Eve represents its beginning for the Lord and his angels did not know about the defection right away. To the intruding Watcher, Eve replied that she could eat of any tree in the Garden except for the Tree of Knowledge lest she die. Thereupon the Watcher replied:

You will not die. For God knows that when you eat of it your eyes will be opened, and you will be like God, knowing good and evil. (Gen. 3:4-5)

Convinced by the Watcher that to learn about what was right and wrong would not kill them, the gardeners "ate" and soon began to realize that they were naked. This symbolized their first step out of ignorance. More importantly, they began to realize that their ignorance was the Starmen's method of keeping them under control, under the Lord's authority. But it was just a feeble step. When they heard the sound of the Lord walking in the garden, they hid themselves from His presence among the trees.

...The Lord God called to the man, and said to him, "Where are you?" And he said, "I heard the sound of thee in the garden and I was afraid, because I was naked; and I hid myself." He said, "Who told you that you were naked? Have you eaten of the tree of which I commanded you not to eat?" (Gen. 3:9-11)

This was a crushing blow to the Starmen's operation and put them in a predicament because they could no longer expect their "created" gardeners to follow along with the program blindly:

Then the Lord God said, "Behold, the man has become like one of us, knowing good and evil; and now, lest he put forth his hand and take also of the tree of life, and eat, and live forever -" therefore the Lord sent him forth from the garden of Eden, to till the ground from which he was taken. (Gen. 3:22-23)

To protect the garden from future intrusions, the Lord had a perimeter guard put on the eastern border to seal off the operation:

He drove out the man; and at the east of the garden he placed the cherubim, and a flaming sword which turned every way, to guard the way to the tree of life. (Gen. 3:24)

This is a fascinating passage - the concept of a *flaming sword* in the times of Adam. Today we would call such a device a remote controlled laser, and the cherubim (plural for cherub), robot sentries. Physical beings, a physical war, and now physical equipment. We know also that Eden was a physical place. A hint as to the conditions in Eden is given by the expelled gardeners in the apocryphal *Book of Adam and Eve.*

When Adam and Eve were driven out of the Eden base, they were not totally foresaken by the Lord for they were only sent out "a little to the east." Another apocryphal book states specifically that they were only forty-five cubits away, or just outside the perimeter. [II Adam 11:8-9] (117) They remain on the "holy mountain," one of many mountain bases alluded to in the Old Testament, and for a time live in the "Cave of Treasures." During their first days out of the Garden, they are in complete shock, and they lament that they no longer enjoy the perpetual light which they once knew:

Oh Eve! Remember that while we were in the garden, we knew neither night nor day... For, so long as we were in the garden, we neither saw nor even knew what darkness is. (I Adam XI:9, XII:9)

This was apparently a physical light, an artificial light within the garden. They were thus kept indoors all of their lives, and it is this quality of "divine" light which so intrigues Adam and Eve that they are described as constantly seeking it. When cast out of the garden, they are introduced to darkness and the light of day and are terrified. But the Lord comforts them:

Strengthen thy heart, and be not afraid. This darkness is not punishment. But O Adam, I have made the day, and have placed the sun in it to give it light. (I Adam XIII:12)

Thus an intriguing strand of evidence points not only to the biological creation of the Hebrews from existing Earth stock by the Starmen, but of a "garden" base, where the created servants worked for the Lord under artificial light and were never permitted outside until a traitorous Watcher seduced Adam and Eve away from their innocence, thus ruining the Lord's plan for the gardeners and necessitating that they be evicted from the compound.

When Adam and his wife are first cast out of the Garden, they are placed into the Cave of Treasures, a kind of transition house where they are taught to survive in the wilderness outside of the Eden compound. During these times Satan comes around to tempt the gardeners further, sometimes manifesting himself in a bright light and pretending to be an angel of the Lord. But the gardeners are carefully monitored, and each time they are rescued as the war appears to escalate.

Eventually Adam and Eve adjust to their new environment and begin to have children. The story of the first two children, Cain and Abel, is well known. The vengeful Cain killed his innocent brother and, because of this savage act, was condemned by the Lord to exile in the land of Nod (which means "wandering"). This place was described in the *Book of Adam and Eve* as being at the base of the holy mountain. The other descendants of Adam were strictly admonished not to go down to mix with the descendants of Cain. Apparently Seth's lineage was still "pure." However, we begin to realize that the sons of Cain are continuing to be led astray by the fallen Watchers, and it is *his* daughters who are being corrupted:

And Seth the elder, tall and good, with a fine soul, and a strong mind, stood at the head of his people; and tended them in innocence, penitence and meekness, and did not allow one of them to go down to Cain's children.

But because of their own purity, they were named "Children of God," and they were with God, instead of the host of angels who fell; for they continued in praises of God, and in singing psalms unto Him, in their cave - the Cave of Treasures. (II Adam 11:3-4)

What does the author mean by "pure"? The gardeners had supposedly already been tainted with the "original sin," so here we see that the word *pure* had two meanings: the adherence to the teachings of the Lord and the biological pureness of the race:

...the daughters of Cain lusted after the sons of Seth like ravenous beasts, and the sons of Seth after the daughters of Cain, until they committed abomination with them...And God was angry with them, and repented of them because they had come down from glory, and they had thereby lost or forsaken their own purity... (II Adam XX:32, 34)

Adam and his pure kin are taught by the Lord various skills, such as agriculture, and they spend time generally toiling outside the Garden. They apparently do not like their new life of labor, however, as Seth and his children remain obsessed with the goings-on inside the Eden base:

But Seth and his children did not like earthly work, but gave themselves to heavenly things; for they had no other thought than praises, doxologies, and psalms unto God.

Therefore did they at all times hear the voices of angels, praising and glorifying God; from within the garden, or when they were sent by God on an errand, or when they were going to heaven. (II Adam XI:6-7)

Meanwhile the sons of Cain were being irreversibly corrupted by the Watchers, who were illegally visiting Earth. Enoch tells how the angel Azazel taught men to make swords, knives and breastplates and gave them other metalsmithing skills. Another, Baragijal, taught his receptive pupils about astrology (astronomy?). Still others taught men such diverse subjects as meteorology, war, writing and even abortion!

And the fifth was named Kasdeja; this is he who showed the children of men all the wicked smitings of spirits and demons and the smitings of the embryo in the womb, that it may pass away... (Enoch LXIX:12)

Finally the archangels discovered who had perpetrated man's fall:

And then Michael, Uriel, Raphael, and Gabriel looked down from heaven and saw much blood being shed upon the earth, and all the lawlessness being wrought upon the earth... Thou seest what Azazel hath done, who hath taught all unrighteousness on earth and revealed the eternal secrets which were preserved in heaven, which men were striving to learn. (Enoch XI:1, 6)

When the defectors were discovered, full scale war ensued, and they eventually lost their right to return to heaven:

...from henceforth you shall not ascend into heaven unto all eternity, and in bonds of the earth the decree has gone forth to bind you for all the days of the world. (Enoch XIV:5)

But the story becomes yet more complicated for the now sentenced Watchers began to make treks up the holy mountain to try to entice the remaining pure children of Adam to come down and join the sons of Cain below, who had by this time become quite numerous. Jared reports an instance when a party of 30 men came to him:

Satan then appeared to him with thirty men of his hosts, in the form of handsome men; Satan himself being the elder and tallest among them, with a fine beard. (II Adam 17:4)

Enoch also describes these beings as apparently no different from ordinary men, although possibly whiter:

And I raised mine eyes again to heaven, and I saw in the vision, and behold there came from heaven beings who were like white men. (Enoch 87:2)

Jared continues:

Then Jared and they exchanged a greeting and he said to the elder among them, "O my father, explain to me who these are with thee; for they look like strange men." (II Adam 17:8)

In other words the deceiver, Satan, the "fallen" Watcher and leader of others, was merely a man, like mortal men, with a beard and white skin.

Despite all of the warnings by the Lord, his children were slowly lost to these fallen Watchers and the clan of Cain until only three righteous families were left on the mountain: those of Methuselah, Lamech and Noah. Enoch was not included because he was privileged to have been taken up into heaven and thus did not die on the Earth. (Gen. 5:22)

During these times Enoch had a very special relationship with the archangels and the Lord for he was taught many things which were not known to the others, and he was commissioned to make note of various things which were pointed out to him and to record them for future generations. These are the things which are recorded in the *Book of Enoch.* Enoch was the first true proxy leader for the budding race of Hebrews, the forerunner to a new method of dealing with the people which the Starmen had created.

Let us pause for a moment to review. We first stated that the Starmen established the Eden base and created the Hebrews to serve them out of existing earth stock in the Mesopotamian region. One of the Watchers, Gadreel, who was supposed to keep watch over the operation, or some part thereof, tempted Eve, who in turn taught Adam about their state of ignorance in the garden. When discovered, the gardeners had to be exiled because they knew too much and could no longer be trusted to obey blindly the commands of the Lord. But they were kept near the perimeter of the compound, looked after and trained to care for themselves in the earth's natural environment.

Several generations later, we are told that Cain kills his brother Abel. For this he was sent farther away, to the base of the holy mountain, where he takes a wife from the existing stocks of naturally evolved humans living in the area and has many children.

The Watchers at this time begin to see that the daughters of men were fair and took wives of them, thereby defiling themselves. For this corruption, war broke out. The Watchers were eventually defeated and subsequently banished to the Earth. They had not only acted irresponsibly in getting intimate with the inferior Cain offspring, but they had taught them too many negative values which would then be handed down from generation to generation and which could ultimately end in the whole world's self-destruction, for there was too much knowledge without the appropriate values to channel it into constructive ways. Let us move on.

The patriarch Enoch was then judiciously selected to learn specific knowledge which was to be passed on to the remnant families on the holy mountain who remained loyal to the Starmen and eventually handed down to future generations. A new order was in the works. A new way of dealing with the righteous remnant had to be devised, a way which would require less direct contact by the Starmen and some psychological incentives to obey. That way was through a "covenant" or moral contract which, if followed, would ensure the Hebrews' establishment on the Earth. It was to be a relationship to their Lord without the direct contact they once knew for all business henceforth would be conducted through proxy leaders.

One of the significant things which Enoch learned was that the Starmen were human like himself and that it was *knowledge* which separated Earthmen from Starmen. These phfsical beings had physical vehicles, and they ascended into heaven in them. Enoch describes something not unlike what has been reported by Earthmen for thousands of years, more recently called UFOs:

And I saw there something like an invisible cloud; for by reason of its depth I could not look over, and I saw a flame of fire blazing brightly, and things like shining mountains circling and sweeping to and fro. (I Enoch CVIII:4)

He not only sees these vehicles, he flies in them and describes one event quite vividly:

And the vision was shown to me thus; Behold, in the vision clouds invited me and a mist summoned me, and the course of the stars and the lightnings sped and hastened me, and the winds in the vision caused me to fly and lifted me upward and bore me into heaven. (Enoch XIV:8)

Enoch is spoken of in the Bible as "walking with God." The term "walking with God" meant a very intimate association between the Earthmen and the Starmen, a face-to-face relationship unknown to most other men. Because Enoch walked with God, he was shown many "visions" or sights, which he was to record. Enoch speaks very matter-of-factly of his flights above the Earth to see these "hidden visions":

And after those days in that place where I had seen all the visions of that which is hidden - for I had been carried off in a whirlwind and they had borne me towards the west - there mine eyes saw all the secret things of heaven that shall be, a mountain of silver, and a mountain of gold, and a mountain of soft metal and a mountain of lead. And I asked the angel who went with me, saying, "What things are these which I have seen in secret?" And he said unto me: "All these things which thou hast seen shall serve the dominion of His Anointed that he may be potent and mighty on the earth." (I Enoch LII:1-4)

The word "vision" today usually means only a psychic phenomenon associated with a religious experience. From these and other passages, it should be apparent to the reader that the term "visions" in this context meant the literal observation of something so fantastic to the observer that it seemed almost dream-like and yet was obviously physically happening. "Visions," therefore, is one of our "words of comparison."

In reference to the mountains of metal described by Enoch, we may never know the reason why the Starmen originally came to the Earth. The tantalizing fragments quoted above, however, entice us to speculate that at least while they were here, they exploited our mineral resources as they saw fit!

Is there any way to know how long ago all of this took place? How long ago, for example, did Eden exist? For an approximation we can use corroborating data from science and Christianity. Scientists claim that there actually was a catastrophic flood in the Mesopotamian region sometime very close to 11,600 years ago (more about this later). By drawing information from the genealogies of Genesis chapters five and six, we can trace back from Noah to Adam to arrive at an approximate date for the existence of Eden. The genealogy of Adam's descendants to Noah is as follows:

When	Adam was	130	years old, his wife bore	Seth
	Seth	105		Enosh
	Enosh	90		Kenan
	Kenan	70		Mahalel
	Mahalel	65		Jared
	Jared	162		Enoch
	Enoch	65		Methuselah
	Methuselah	187		Lamech
	Lamech	182		Noah
	Noah	600	- the flood came upon the earth	
TOTAL		1,656		

By adding the two figures, 1,656 and 11,600, we arrive at the approximate date of 13,256 years ago. Such a base probably existed from a little before the time of Adam until the time of Noah, perhaps 1,500 years. But is there any reason to believe the dates given in the Bible? As argued earlier, a race with a superior technology would surely have no problems in attaining the extended lifespans. Indeed we are on the verge of such a breakthrough.

But the idea behind the Bible itself, especially the Old Testament, is the tracing of a particular lineage of people, the Hebrew people, as a great family. The genealogies are not merely given in passing. Careful attention was paid to include ages of parents at the birth of their children, meanings behind the children's names (i.e. "Peleg," which means "divided"), and the tracing of the lineage from father to son (a very male dominated society), thus tying all of the major patriarchs to the great family tree. Meticulous details such as these were constantly observed in the Old Testament genealogies. Another important argument for the seemingly incredible ages reported is that they are presented as fact:

One would think that if Genesis were really written in some late period of Jewish history, as critics allege, the writer would have interjected some explanatory comment or at least some expression of wonder at the uniqueness of the phenomena he was describing.

But instead he wrote in the most simple and straightforward way possible, as a sober historian or news reporter would do, with no attempt to justify or explain events which would seem almost incredible to later generations. (118)

So we can place a high degree of confidence in the biblical view that the ancient Hebrews once lived multiple hundreds of years. But what about the great flood? Genesis says that there once was a great flood that came upon the Earth in Noah's time because of the vile conditions which had fermented on the earth as a result of the Watchers' transgressions. How did this come about?

After learning the "secrets," Enoch is then returned to his family to declare them to his people and to record them that they might be handed down to future generations:

And those seven holy ones brought me and placed me on the earth before the door of my house, and said to me: "Declare everything to thy son Methuselah, and show to all thy children that no flesh is righteous in the sight of the Lord, for he is their Creator. One year we will leave thee with thy son, till thou givest thy last commands, that thou mayest teach thy children and record it for them, and testify to all thy children; and in thy second year they shall take thee from their midst." (Enoch LXXXI:5-6)

When asked by Methuselah what was to be, Enoch replied:

The Lord will do a new thing on the earth, and this I have already seen in a vision, and make known to thee that in the generation of my father Jared, some of the angels of heaven trangressed the word of the Lord. And behold they commit sin and transgress the law, and have united themselves with women and commit sin with them. And they shall produce on the earth giants not according to the spirit, but according to the flesh, and there shall be great punishment on the earth, and the earth shall be cleansed from all impurity. Yea there shall come a great destruction over the whole earth, and there shall be a deluge and a great destruction for one year. (Enoch CVI:13-15)

With Enoch's great mission completed, he was removed from the earth to live in heaven with the Starmen. All that remains of this in the Bible is a short paragraph in Genesis which concludes: "Enoch walked with God and was not, for God took him." (Gen. 5:24)

The second important proxy leader was to be Methuselah's grandson. This was Lamech's son, Noah. According to the *Book of Enoch,* Noah was a very special child indeed for upon his birth, his father instantly realized that Noah was somehow different:

And after some days my son Methuselah took a wife for his son Lamech, and she became pregnant by him and bore him a son. And his body was white as snow and red as the blooming of a rose, and the hair of his head and his long locks were white as wool, and his eyes beautiful...And his father Lamech was afraid of him and fled, and came to his father Methuselah. And he said unto him: "I have begotten a strange son, diverse from and unlike man, and resembling the sons of the God of heaven; and his nature is different, and he is not like us, and his eyes are as the rays of the sun, and his continence is glorious. And it seems that he is not sprung from me but from the angels..." (Enoch CVI:1-6)

Lamech is reassured that the child is indeed his son; however, Lamech's reaction to the infant is suggestive of another answer: He was *half Starman,* as was often the case in the old days. This matter will be covered in more detail later.

He is also told that the seed of Noah will be preserved in the impending deluge which will come upon the earth to heal and cleanse it. And why will Noah be preserved? As a clue we read in Enoch:

And now my Lord, destroy from the earth the flesh which has aroused thy wrath, but the flesh of righteousness and uprightness establish as a plant of thy eternal seed... (I Enoch LXXXIV:6)

These concepts of the "eternal seed" and the "pure vine" comprise a key theme which will recur throughout the rest of the Old Testament. The "righteous branch" of the "pure vine" shall be established on the Earth as a legacy of the Starmen. But what *is* the "eternal seed"? It is sperm, carried by the lineage of Starmen from however long ago that they began to conquer space and spread it. It is now received by Earth women and born unto them as the "mighty men of old, the men of renown."

Recall that the *central* theme of the Old Testament is the establishment of the Hebrews on Earth, even at the cost of displacing and killing other races such as the Canaanites and the Hittites! These created men were not of the purely evolved stock found on the Earth; they were a hybrid variant developed, "created," by the Starmen during their mission on the Earth. The Earth therefore, was "planted" with the "seed" of the Starmen. Noah was such a hybrid who became a famous proxy leader for the Starmen on their mission on Earth. This idea will be reinforced as other evidence is presented in a subsequent section.

The Flood

The flood legend, like the Eden Metaphor, is an account from the dimmest history of an ancient culture. The final version of the biblical flood story was compiled by an unknown author from three different sources: the *J Document,* the *P Document,* and the *S Document.* Each source lends its own particular perspective to the account by providing details not mentioned by the others and by reflecting the authors' various political and theological orientations. The legend has proven to be extremely adaptable in its different forms, but despite its evolution from earlier Sumerian and Babylonian versions (from which it is thought to have originated), the basic message has remained intact: there was a great flood, a righteous man was saved from it, and his "seed" was preserved to repopulate the Earth.

As stated previously, the biblical version displays some inconsistency due to the latest compiler's apparent confusion about the story's proper contents, which created contradictions in some places. For example Noah's sending of birds to discover if the waters were subsiding and his sacrificial offering are from the *Jahovist* or *J Document.* The detailed instructions for the building of the ark and the rainbow of the covenant are peculiar to the *Priestly Code* or *P Document.* In J the flood culminated in forty days; in P it takes 150. In the *J Document* the animals take a week to embark; in P, apparently only a day. One version has Noah bringing two of every sort of animal (Gen. 6:19) while the other specifies that he bring only "seven pairs of clean animals ...and one pair that is not." (Gen. 7:2)

That there are so many cultures around the world having flood legends leads some to believe that there may indeed have been a flood of global proportions at some remote period in the past. There is abundant scientific evidence for such a flood too. The last ice age began about 30,000 years ago and reached its peak about 18,000 years ago. Then, as the earth's climate again returned to a warmer cycle, the great ice caps which had been advancing over the northern continents for thousands of years suddenly and catastrophically collapsed, raising the world's oceans by 300 feet.

"About 11,600 years ago, the North American ice cap underwent a sudden collapse, followed by rapid melting," according to [Cesare] Emiliani. "A huge amount of ice-melt water rushed into the Gulf of Mexico and produced a sea-level rise that spread around the world with the speed of a great tidal wave, which can circle the globe in 24 hours." (119)

According to Emiliani and others, the date of the flood can be determined by studying core samples drawn from the ocean floor. A tiny organism called a *foraminifer* chemically locks into its shell a record of the temperature and salinity of the ocean. When these surface-dwelling organisms reproduce, they shed their shells, which then fall to the ocean floor and accumulate over the aeons in layer after layer. Such accumulations may go back as far as 100 million years. Radio-carbon dating and other methods can be used to determine the ages of various levels of sediment, providing oceanographers with an accurate record of the ocean's salinity for any given period. It was this marked salinity decrease in the layers of tiny foraminifera which thus signaled a global flood originating from North American ice caps some 11,600 years ago.

Otto Heinrich Muck believes that the legend of the great flood was the result of a huge asteroid which collided with the Earth, striking parts of the Carolinas and gouging out the undersea trough called the Puerto Rico Trench. The impact of the asteroid at this critical point in the earth's crust triggered violent undersea earthquakes which sent tidal waves around the globe some 11,000 to 10,000 years ago. (120) Perhaps Emiliani's evidence supports Muck's theory and the two events are related. Some theorists claim that the flood story can be explained in terms of several localized floods which occurred throughout history in various areas of the world, accounts of which were then handed down from generation to generation, gradually assuming the proportions of a global event.

In 1922 the eminent British archaeologist Sir Leonard Woolley discovered in a large Iraqi mound or "tell" (once part of the great Sumerian city of Ur) the first concrete evidence of this biblical deluge. While excavating the Royal Cemetery, workers suddenly came upon what appeared to be virgin soil.

What perplexed Woolley was the sudden disappearance of any trace of culture or habitation which surely preceded the highly sophisticated one just uncovered.

Woolley knew that great civilizations do not simply spring into existence but build upon the rubble of previous cultures, and so he continued to dig through eight feet, two inches of silt and clay. Finally his persistence paid off dramatically for below the silt he discovered flint implements and fragments of pottery made by a late Stone Age culture. But the entire historical record of the intermediary habitations had obviously been swept away in a deluge large enough to have left sediment eight feet deep!

Archaeology has made it clear that the Mesopotamian valley and delta regions were especially vulnerable to such floods due to the funnel effect produced by their topography as it faces the sea. (121) A flood of the proportions described by Emiliani or Muck would thus have had disastrous consequences for this region especially. If the ice-melt described by Emiliani, for example, sent a tidal wave around the globe, its force would have been magnified many times over as ocean water surged up the gulf and into the narrowing Tigris-Euphrates delta region. So the flood described in the Bible was more than merely a rapid rise in the water level: it was a devastating surge from the Persian Gulf.

We would also expect that this period of rapidly changing climatic conditions was marked by torrential downpours. Water from Emiliani's proposed ice-melt flowing across warming continents would also likely produce great quantities of water laden clouds, and it was just such a combination of massive ice-melt, funnel-like topography and water laden clouds which may have produced the disastrous Mesopotamian flood that inspired the biblical account.

So, whatever the sequence of events, it is clear that the Bible and science are in agreement over still another point: a terrible flood really did occur in Mesopotamia. It would be useless to speculate whether the Starmen were responsible for causing the flood or whether they merely knew that some natural disaster was imminent. But the *Book of Enoch* tells us that the Starmen knew of the impending disaster at least at the time of the birth of Noah. And if we take the Bible chronology at face value, then they knew at least 600 years in advance.

When the Starmen realized that such a scenario was developing, they set about to prepare the created men. Detailed instructions were given to make an ark which would preserve Noah and his families. The biblical cubit is 17.5 inches long. According to the plan, the three-storied ark was to be 300 cubits or 437.5 feet long, 73 feet wide and 44 feet high - every bit as large as a modern cargo ship!

Genesis 6:14 says that Noah was to make an ark, but the *Book of Enoch* says that it was the angels who made it for him:

And in those days the word of God came unto me, and He said unto me: "Noah, thy lot has come up before me, a lot without blame, a lot of love and uprightness. And now the angels are making a wooden [building], and when they have completed the task I will place my hand upon it and preserve it, and there shall come forth from it the seed of life, and a change shall set in so that the earth will not remain with out inhabitants." (I Enoch LXVII:1-2)

Noah then faithfully collected pairs of creatures to be saved, and he and his families were shut in by the Lord, for the time had come:

In the six hundredth year of Noah's life, in the second month, on that day all the foundations of the great deep burst forth, and the windows of the heavens were opened. And rain fell upon the earth forty days and forty nights. (Gen. 7:11-12)

There was not only a long period of rain, there was a tidal wave from the "great deep," or the Persian Gulf. To further the notion that such a tidal wave came from the Gulf, we read that the ark came to rest in mountains of Ararat, which lie far north of the delta, possibly as far as 800 miles, near the source of the Great Zab River in Turkey.

After nearly five months, the waters abated and Noah came forth from the ark. God then made an agreement with Noah, a covenant, saying that he would establish his children on the Earth:

Then God said to Noah and to his sons with him, "Behold, I establish my covenant with you and your descendants after you..." (Gen. 9:8)

The flood may have been a global event, but its impact worldwide was nowhere near as great as in the Mesopotamian valley region. A mere 300-foot rise in the sea level would have sent coastal dwellers for high ground, but the majority of normally evolved humans had only survived yet another test of nature. Meanwhile the special "seed" of Noah had been preserved in the great ark. After the flood Noah and his families were released into the valley and instructed to "go forth, be fruitful and multiply." The Old Order thus ended.

In looking back at the events portrayed in the Bible up to this point, an observation can now be made regarding the portrayal of the biblical characters. In metaphoric form, Genesis chapters nine and ten describe how the sons of Noah - Shem, Ham and Japeth - went forth to repopulate the world (their world, the Mesopotamian valley). But Shem, Ham and Japeth are not real people according to some sources. They are thought to represent the three major Middle Eastern races: the Semites, from which the Hebrews arose, the Hamites and the Aryan or Indo-european peoples. (122)

Likewise in the Eden metaphor the Hebrew word "adam" represented not an individual, but "man" in the sense of "mankind." Only much later was "adam" (as was "satan") reduced to a proper name. So in interpreting the story of Eden, it was probably true that Adam and Eve represented their entire clan of gardeners. As in the case with Noah's offspring, the characterization of an entire race was thus compacted into one entity, and an entire history into one name. Because of the extremely long time spans being dealt with in the Old Testament, such cursory portrayals are to be expected, especially among the minor characters.

But Noah himself, it must be argued, was a real person. Too many details about his life are depicted even if many possibly erroneous conceptions later evolved around him. The integrity of the character is maintained throughout the text of the flood story so as to leave little doubt that there was once a man, be he Noah, Uta-napishtim or Ziudsuddu, who was saved from the great flood by the Lord.

These examples point out the evolving format of biblical characters. The lineage of the race is carefully traced, but in the process the less significant characters became more compressed over time into groups represented by one name while the more important characters generally survived intact.

It must be emphasized that the stories in the Old Testament up to perhaps the time of Abraham were handed down from generation to generation by word of mouth, and only with the invention of writing were the stories finally recorded. If the flood occurred 12,000 years ago, and writing was developed even as early as 6,000 years ago, (123), there would have been as much as 6,000 years for the evolution of these stories. But whether Noah's sons were real individuals or simply eponymic characterizations of lineages of people is not important for the thesis of this book. What is important is the idea that Starmen, represented as angels, were concerned with a righteous Noah and his "seed," and ensured their survival for the establishment of the race on Earth.

Interestingly, it is during the period right after the flood, about 10,000 years ago, that anthropologists credit man with his remarkable rise in culture and civilization in this region. The chief catalyst for man's conversion from wandering hunter-gatherer modes to the sedentary social mode, they believe, was agriculture:

> *And there can be no doubt that the sharpest and most dramatic shift of gear in our ancestors' progress along the path of human evolution was the invention of agriculture ten thousand years ago.* (124)

"Noah was the first tiller of the soil," says Genesis 9:20. In keeping with the Eden Metaphor, it becomes rather obvious that Noah and his families knew agriculture, which they learned from God. (Isaiah 28:23-29) It makes sense, then, that as the normally evolved men began to migrate back into the Mesopotamian region after the flood, they acquired the knowledge of agriculture from the hybrid Hebrews. Eventually the "agricultural revolution" generated the population explosion, which fermented into the beginnings of the first great Mesopotamian civilizations some 6,000 years later in 4,000 B.C.

Endnotes

109. Leakey, *op. cit.,* p. 92.

110. Zecharia Sitchin, *The Twelfth Planet,* Avon Books, New York, 1976, p. 4.

111. Samuel Noah Kramer and Editors, *Cradle of Civilization,* Time-Life Books, New York, 1967, p. 17.

112. Dora Jane Hamblin and Editors, *The First Cities,* Time-Life Books, Inc., New York, 1975, p. 58.

113. A. Cranston, "Coming; Longer Life Spans," *Reader's Digest,* January 1979, p. 79.

114. *Ibid.*

115. Jean Sendy, *Those Gods Who Made Heaven & Earth,* Berkley Medallion Books, New York, 1972, p. 141.

116. Sitchin, *op. cit.,* p. vii.

117. Platt, Jr., *op. cit.,* II Adam 11:8-9, p. 68.

118. Henry M. Morris, *The Bible Has the Answer,* Baker Book House, Grand Rapids, MI, 1975, pp. 100-101.

119. F. Warshofsky, "Noah, the Flood, the Facts," *Reader's Digest,* September 1977, p. 133.

120. Thomas Fleming, "Solving the Lost-Continent Mystery," *Reader's Digest,* August, 1978, p. 132.

121. Madeline S. and J. Lane Miller, *Harper's Bible Dictionary,* Harper & Row, Inc., San Francisco, 1973, p. 198.

122. *Ibid.,* p. 495.

123. *Ibid.,* p. 828.

124. Leakey, *op. cit.,* p. 123.

II Samuel 22:7-16

Chapter IX
THE INTERMEDIATE ORDER DEFINED

We have heard with our ears, O God
Our fathers have told us,
what deeds thou didst perform in their days,
in the days of old:
thou with thy own hand didst drive out the nations
but them thou didst plant.
— Psalms 44:1-2

The Intermediate Order comprises the major portion of the Old Testament from approximately the time of Abram, (Abraham; 20th -19th century B.C.) to the period just before the arrival of Christ (c. 490 B.C.). During these times we are dealing with real, historical people. The Starmen now conduct their operation in a different manner than during the Old Order: all exposure to the masses of Hebrews is minimized, and most transactions are carried out through proxies.

This portion of the Old Testament was written largely from eyewitness accounts or from sources not far removed from the actual events. This contrasts with the Old Order, in which most of the characters are eponymic representations of entire races or traditions and the sequence of events remote from the authors and obscure.

Basically it is through this segment that the Starman Intervention Theory stands or falls. It is here that we find the most convincing evidence of Starman activity. From this portion of the Old Testament we should be able to visualize not only their appearances, their characters, their vehicles and their equipment, but their methods and purpose as well. If these things can be brought out through the scriptures themselves, then it will become clear why the Hebrews were so consistent in the recording of their religion and the worship of their god and also understand why the Old Testament had such a profound impact on subsequent generations of Hebrews and Gentiles alike.

The Lord

The term "God" did not always represent a single entity.The form worshiped from the time of Adam and Noah and possibly to Abraham was "elohim," plural for *gods.* Later editors of the texts changed the form to a proper name, "El" or "God," just as Adam's and Satan's names were changed from their plural or generic forms - adam (mankind) and satan (deceiver). The significance of this, of course, is that the Starmen were originally known as a group of gods, as the Old Order segment of the Old Testament has shown. Interestingly a few of these old style pluralities remain to this day in the Old Testament despite the screening. Genesis 3:22 reads: *"behold, the man has become like one of us, knowing good and evil."* The famous Tower of Babel incident is another good example of this plurality:

And the Lord came down to see the city and the tower which the sons of men had built. And the Lord said, "Behold, they are one people, and they have all one language; and this is only the beginning of what they will do; and nothing that they propose to do will now be impossible for them.

Come, let us go down there and confuse their language, that they may not understand one another's speech." (Gen. 11:5-7)

This plural form changes with the Intermediate Order to El Shaddai, or "God of the Mountains." The root of the name El Shaddai, according to *Harper's Bible Dictionary* (125) is "to act violently." Both of these names are quite appropriate to the *true* nature of the Starmen's character, for their vehicles often produced violent effects, such as high winds, loud noises, flashes of light, etc., as will be shown, and they maintained bases on various mountains away from the masses:

Sing praises to the Lord, who dwells in Zion! (Psalms 9:11)

Great is the Lord and greatly to be praised in the city of our God! His holy mountain, beautiful in elevation, is the joy of all the earth, Mount Zion, in the far north... Walk about Zion, go round about her, number her towers, consider her ramparts, go through her citadels; that you may tell the next generation that this is God, our God for ever and ever. (Psalms 48:1-14)

In Judah God is known, his name is great in Israel. His abode has been established in Salem, his dwelling place in Zion. (Psalms 76:1-2)

The Lord came from [Mount] Sinai, and dawned from [Mount] Seir upon us; he shown forth from Mount Paran, he came from the ten thousands of holy ones, with flaming fire at his right hand. (Deuteronomy 33:2)

"For on my holy mountain, the mountain height of Israel," says the Lord God, "there all of the house of Israel, all of them, shall serve me in the land." (Ezekiel 20:40)

In the twenty-fifth year of our exile, at the beginning of the year, on the tenth day of the month, in the fourteenth year after the city was conquered, on that very day, the hand of the Lord was upon me, and brought me in the visions of God into the land of Israel, and set me down upon a very high mountain, on which was a structure like a city opposite me. (Ezekiel 40:1-2)

The evolution of the concept of God from Elohim to El Shaddai represents a fundamental change in the relationship between the Starmen and their people. As stated before, in the Eden Metaphor God was a physical being, a rather paternal figure who walked among men and spoke to them face to face:

And they heard the sound of the Lord God walking in the garden in the cool of the day, and the man and his wife hid themselves from the presence of the Lord God among the trees of the garden. But the Lord God called to the man, and said to him, "Where are you?" (Gen. 3:8-9)

In the Intermediate Order, God has become aloof from his people. He now lives in the "high places," in the hills and mountains away from the masses. This is not to say that God is no longer a physical being; it is just that the mode of operation has changed. And while He or His angels are still seen from time to time by key biblical characters, His exposure is greatly diminished. The Bible makes it clear that He still has human form however:

"...and I will walk among you, and will be your God." (Leviticus 26:12)

"...Thou, O Lord, art in the midst of this people; for thou, O Lord, art seen face to face..." (Numbers 14:14)

"Because the Lord, your God walks in the midst of your camp, to serve you and give up your enemies before you, therefore your camp must be holy, that he may not see anything indecent among you, and turn away from you." (Deuteronomy 23:14)

Often when God comes to his people, He now does so from "heaven" or his holy mountain, and His arrivals are always marked by terrible and magnificent displays which never cease to strike fear and awe into the hearts of all who witness them. It is during this time that God begins to reveal His word through selected representatives from among the Hebrews, and it is here that the concept of God probably became so distorted.

The reason for this distortion is plain to see if we can picture mission of Starmen in a primitive Hebrew environment. The leader would be the "Most High," and his subordinates would be angels. The title "Lord God" distinguished the Lord from his subordinates, who were also seen as gods. As the angels had names, so too did God, according to the psuedepigraphal Enoch II: *"On the tenth heaven is God, in the Hebrew tongue he is called Aravat."* (II Enoch XX:3) This is to be differentiated from his title as Lord, or as God, or even as YHWH, which probably meant "to be."

Similarly the Starmen and their equipment are consistently misidentified throughout the Old Testament. They flew inside their spaceships, but the ships were not God, as is generally portrayed by theologians today. The things with which God eventually became equated were clearly not God "the leader," or an angel, but things which were associated with Him and only later were combined to form the composite we know today. God was *not* the vehicles (clouds, pillar of fire, etc.) in which He flew, He was *not* the special equipment (cherubim, ark, etc.) He employed, and He was *not* the stupefying forces (spirit, holy things, etc.) which He had at his disposal.

Furthermore, God's manifestation is marked by a much more stern and demanding attitude toward His people during this period. His plan for the Hebrews now has a sense of urgency about it that didn't seem to exist in the Old Order, and He permits little or no deviation. This is reflected in the healthy respect for His power by His Hebrew hosts:

Serve the Lord with fear and trembling. Kiss his feet lest he be angry, and you perish by the way; for his wrath is quickly kindled. (Ps. 2:11)

The friendship of the Lord is for those who fear him... (Ps. 25:14)

The human nature of the Lord, however, is perhaps best understood through the logic of the Bible stories. Three important points can be made concerning God's characterization: First, God was portrayed throughout the Old Testament as having little patience with his people. Several times during the exodus, He is brought to the point of annihilating the entire race. Fortunately in each instance He is moved by Moses to have mercy and relents. (Ex. 32:9-14; Num. 14:11-20)

A similar sentiment arises during the time of David (I Chron. 21:1-27), but God is again swayed to have mercy on the Hebrews. This points to a second human trait and that is the necessity for God changing His mind from thoughts of annihilating the Hebrews to thoughts of their preservation. He was not the steadfast spirit God as is preached today, but a sometimes vacillating, indignant general of the Starmen. Finally we begin to realize that God's knowledge, as portrayed in the Old Testament, is not perfect. A strange paradox arises in Psalms:

Thou knowest when I sit down and when I rise up; Thou discernest my thoughts from afar...and art acquainted with all my ways. Even before a word is on my tongue, lo, O Lord thou knowest it altogether. (Ps. 139:2)

God must have known therefore that His angels would defect in the Garden and lead Eve astray. But in fact, he did not, for as God is walking through the Garden he calls to the gardeners;

"Where are you?"... "Have you eaten of the tree of which I commanded you not to eat?"... "Who told you that you were naked?" (Gen. 3:9, 11)

This human nature of God pervades the Old Testament and leaves little doubt as to God's misidentification. His temperament, knowledge and consistency indicate human attributes because He was a Starman.

Angels

Mention "angels" to the average person today, and the first thing that probably comes to mind is the classical ministering spirit with white robes and wings. The word "angel" means "messenger," but not all things lumped into the category of "angels" fit into the mold of spirit beings who brought messages.

There are primarily three manifestations into which angels have been catagorized. First there are the *cherubim* (plural for cherub). But these are more suggestive of physical, *mechanical* devices, not spiritual beings as portrayed today. For example the cherubim which were placed at the entrance to the Garden of Eden with a "flaming sword" to guard the way to the Tree of Life (Gen. 3:24) were clearly understood to be different from angels even at the time the scripture was written. There were also the cherubim which stood above the Ark of the Covenant (Exodus 25:18-22), which itself was a mechanical device used to communicate with God. Whenever the Israelites got into trouble they went to the Ark and called upon God, who answered from "between the cherubim" which were affixed to the lid of the Ark. There was also the cherub in which the Starmen flew:

He bowed the heavens and came down; thick darkness was under his feet. He rode on a cherub and flew. (II Samuel 22:11-12)

Clearly the cherubim were mechanical devices employed by the Starmen, but not the Starmen themselves. Seraphim and ophannin are also mentioned in this context in the Bible. Isaiah chapter six describes seraphim as having six wings and as being able to fly. Other brief mentions of the cherubim, seraphim and ophannin together in the *Book of Enoch* (I Enoch LXI:10) suggest that these were all understood as devices of some sort.

The second form to which the term "angels" has historically been assigned is the "angel of the Lord" characterization. There are two types of things represented in this concept, the first being the Starmen and the second the vehicles in which the Starmen flew. Moses first discovers God's presence when he comes upon the "burning bush" on Mount Horeb:

And the angel of the Lord appeared to him in a flame of fire out of the midst of a burning bush...(Exodus 3:2)

And yet it was *God* who spoke to him from the "angel" or the ship. Another reference to "angel of God" in the context of "spaceship" is during the exodus. As the Israelites are leaving Egypt, a mysterious "pillar of cloud" or "pillar of fire" goes before them to lead them. But the reference to this "pillar" shifts from God to pillar to angel of God quite interchangeably:

And the Lord went before them by day in a pillar of cloud to lead them along the way...(Ex.13:21)

Then the angel of God who went before the host of Israel moved and went behind them; and the pillar of cloud moved before them and stood behind them...(Ex. 14:19)

Often this "angel of the Lord" was not a being, but the ship in which the Starmen flew. A passage in Exodus clarifies this by explicitly stating that the Lord flew the vehicle from *within:*

And in the morning watch the Lord in the pillar of fire and of cloud looked down upon the host of the Egyptians, binding their chariot wheels so that they drove heavily...(Ex.14:24-25)

We often run into this confused image of the manifestation of God because the beings who flew in the vehicles were always associated with them and therefore often grammatically equated with them. This has given rise to some awkward portrayals of angels:

David looked up and saw the angel of the Lord standing between earth and heaven. (I Corinthians 21:16)

In other words he saw the vehicle, which he knew was associated with the angels of God, in the sky just as one might say "John saw Jill on the freeway," implying that Jill was driving in her vehicle on the freeway.

One more such example should be sufficient to illustrate this point. After Jesus was placed in the tomb after his crucifixion, an "angel of the Lord" descends to claim him:

Now after the sabbath, toward the dawn of the first day of the week, Mary Magdalen and the other Mary went to see the sepulcher. And behold, there was a great earthquake; for an angel of the Lord descended from heaven and came and rolled back the stone, and sat upon it. His appearance was like lighting, and his raiment white as snow. And for fear of him the guards trembled and became like dead men. (Matthew 28:1-4)

Although described as a "he," and as having spoken to the two women standing nearby, a strong argument can be made for the idea that this was a Starman in a vehicle. First there is the sensation of an earthquake - a common sensation near rockets and big planes today. There is the descending from "heaven," or the sky, and the blindingly bright light - today recognized as a landing light as the vehicle maneuvers to a landing on top of the tomb. Next there is the voice which came from the vehicle, as over a loudspeaker - a common arrangement on modern police helicopters as they address people on the ground. And finally there is the overall impression the whole scene made on the observers, especially the guards, who were struck dumb with fear. This is in keeping with the presence of this Lord throughout the Bible whenever He manifests Himself to Earthmen. In fact there is a very good consistency throughout the Bible to these types of portrayals of angels and their equipment, thus greatly strengthening the argument that they are mislabeled.

The third category represented by the term "angel" is the human form, the Starmen themselves. Usually the angels were thought to be human even if they were different in some minor respects such as skin color, perhaps slightly different build, features, etc., and technological superiority. The angels of the Old Testament were known to eat, drink and sleep just as humans and were often called simply "men." One time the patriarch Abram (Abraham) was sitting at the door of his tent in the heat of the day, "and behold, three men stood before him." Abraham said:

"My Lord, if I have found favor in your sight, do not pass by your servant. Let a little water be brought and wash your feet, and rest yourselves under the tree..." Then he took curds, and milk, and the calf which he had prepared, and set it before them; and he stood by them under the tree while they ate. (Gen. 18:1-9)

On the eve of the destruction of Sodom and Gomorrah, two angels came to the gate of Sodom and there found Lot, Abraham's nephew. After some persuasion the angels came to his house:

But he urged them strongly; so they turned aside to him and entered his house; and he made them a feast, and baked unleavened bread, and they ate. (Gen. 19:1-3)

If the angels were portrayed as "greater than all the (pagan) gods" or "terrible in glorious deeds and wonders," it was not due to bodily strength, but to the strength of their technology. So "normal" were they that Jacob wrestled with an angel but did not realize it until after they had fought!

And Jacob was left alone; and a man wrestled with him until the breaking of the day. When the man saw that he did not prevail against Jacob, he touched the hollow of his thigh; and Jacob's thigh was put out of joint as he wrestled with him. Then he said, "Let me go, for the day is breaking."

But Jacob replied,

"I will not let you go unless you bless me." Then he said, "Your name shall no more be called Jacob, but Israel, for you have striven with God and with men, and have prevailed." (Gen. 32:24-28)

It appears that the people of Israel understood that the angels of God were human but were superior in some ways which they did not always comprehend:

For the Lord will rise up as on Mount Perazim, he will be wroth as in the valley of Gibeon; to do his deed - strange is his deed!; and to work his work - alien is his work! (Isaiah 28:21)

Vehicles

The Israelites were a small race without a country in the beginning. But the Lord promised his people that they would someday have a country of their own. As the Israelites moved about from place to place in their early days, they came upon other empires and other kingdoms and fought many battles. But the Israelites, though few in number, had the singular advantage of having the Starmen as their allies, and they were "on call" whenever Israel needed help. In a song sung by David after a battle with the Philistines at Gob, we find an excellent example of such assistance and how their vehicles were employed toward this end:

In my distress I called upon the Lord; to my God I cried for help. From his temple he heard my voice, and my cry to him reached his ears.

Then the earth reeled and rocked; the foundations also of the mountains trembled and quaked, because he was angry.

Smoke went up from his nostrils and a devouring fire from his mouth; glowing coals flamed forth from him.

He bowed the heavens and came down; thick darkness was under his feet. He rode on a cherub and flew; he came swiftly upon the wings of the wind.

He made darkness his covering around him, his canopy thick clouds dark with water. Out of the brightness before him there broke through his clouds hailstones and coals of fire.

The Lord also thundered in the heavens, and the Most High uttered his voice, hailstones and coals of fire. And he sent forth his arrows and scattered them; he flashed forth lightnings and routed them. (II Samuel 22:7-15)

Though this event was described through the pen of a man who lived perhaps 3,000 years ago, we can still understand that he was graphically describing a vehicle which flew, presumably by rockets or some similar propulsion device, and shot laser beams or some form of weapon discharging bolts of "lightning." More importantly, the narrator understood that the Lord flies on (or *in*) a cherub, and was separate and distinct from it.

Even the sequence of events was correct: first the thunder of the firing rockets and the associated trembling of the ground, then the smoke and the fire from the engine nozzles (nostrils), then the vehicle rising up and arching to the battle area from the sanctuary, then the clouds and thunder from the descent, and finally, as the vehicle hovers above the scene, the destructive "lightnings" sent forth to rout the enemy! There can be little doubt about what transpired here, and we shall see there are other, equally impressive accounts in the Bible.

The vehicles of the Starmen were usually described as clouds, but other metaphors were also used. Often the vehicles were simply identified as God Himself:

So the Lord of Hosts will come down to fight upon Mount Zion and upon its hill. Like birds hovering, so the Lord of Hosts will protect Jerusalem. (Isaiah 31:4-5)

Particularly vivid descriptions from the visions of Ezekiel and during the early phases of the exodus depicted the presence of the Lord shrouded in thick dark clouds. But in almost all other accounts of encounters with the Lord as well these "clouds" are associated with the activities of the Lord:

And the Lord went before them by day in a pillar of a cloud, to lead them along the way, and by night a pillar of fire to give them light, that they might travel by day and by night; the pillar of cloud by day and pillar of fire by night did not depart from before the people. (Exodus 13:21-22)

But let us emphasize that *even the ancients understood* the Lord *only flew* the vehicle which produced the cloud for the cloud and the Lord were separate:

Bless the Lord...who makest the clouds thy chariot, who ridest on the wings of the wind...(Psalms 104:1, 3)

Then Moses went up on the mountain, and the cloud covered the mountain. The glory of the Lord settled on Mount Sinai, and the cloud covered it six days; and on the seventh day he called to Moses out of the midst of the cloud. Now the appearance of the glory of the Lord was like a devouring fire on top of the mountain in the sight of the people of Israel. And Moses entered the cloud, and went up on the mountain. And Moses was on the mountain forty days and forty nights. (Exodus 24: 15-18)

Notice the reference to the "glory of the Lord." This too is a consistent metaphor for the physical effects associated with the vehicle: its sound, the trembling earth, the high-intensity light and the fiery exhaust blast of the rockets. The cloud settled on the mountain and the voice called from the midst of it, just as in the case of the resurrection of Jesus.

The exhaust of the rockets and their associated thunder was particularly terrifying to the primitive man and was often referred to as God Himself: *"For the Lord your God is a devouring fire, a jealous God."* (Deuteronomy 4:24) By whatever means the vehicles flew, they stirred up a tempest wherever they passed close to the earth:

And behold the Lord passed by, and a great and strong wind rent the mountains, and broke in pieces the rocks before the Lord, but the Lord was not in the wind; and after the wind an earthquake, but the Lord was not in the earthquake; and after the earthquake a fire, but the Lord was not in the fire; and after the fire a still small voice. And when Elijah heard it he wrapped his face in his mantle and went out and stood at the entrance of the cave. And behold, there came a voice to him and said, "What are you doing here Elijah?" (I Kings 19:11-13)

Job understood the distinction between the vehicle and its controller when he declared: *"He covers the face of his throne and spreads over it his cloud."* (Job 26:9) And the Bible is full of verses describing these vehicles. Psalms says it poetically:

Lift up a song to him who rides upon the clouds; his name is the Lord...(Ps. 68:4)

God has gone up with a shout, the Lord with the sound of a trumpet. (Ps. 47:5)

One other form used to suggest these vehicles, or at least some phenomenon associated with them, is the dove. For example Christ had just been baptized by John the Baptist (to signal the commencement of his ministry) when we are told that he went up immediately from the river:

"...and behold, the heavens were opened, and he saw the spirit of God descending like a dove, and alighting on him." (Matthew 3:16)

The "spirit" of God was not a dove, but descending *like* one. It did not produce the rending, tumultuous storm conditions associated with other accounts of vehicles, but it is possible that a vehicle remained high above the scene, its general shape and color prompting the observer to compare it to a dove. To support such a notion, we can draw upon the next verse, which reads:

...and lo, a voice from heaven, saying, "This is my beloved Son, with whom I am well pleased."

A Starman in his vehicle stationed above the scene thus introduced Jesus at the beginning of his ministry. Isaiah 60:8, which reads, *"Who are these who fly like a cloud and like doves to their windows?"*, seems to confirm that clouds and doves are the metaphors used to describe the alien vehicles.

Drawing on the only experience they had, that of such natural forms as clouds and doves, the ancients sought to convey the message that a thing they knew to be neither a cloud nor a dove but something quite undefinable was plying the skies during those times:

Behold, the Lord is riding on a swift cloud... (Isaiah 19:1)

Eventually the Starmen also removed the savior Jesus into heaven in a cloud-enveloped vehicle:

...as they were looking on, [Jesus] was lifted up, and a cloud took him out of their sight...(Acts 1:9)

One final analogy can be made to strengthen the argument that the Lord and His vehicle have been mistakenly lumped together as one phenomenon. Philip, a contemporary of Paul and an evangelist for the budding Christian religion, was told one day to head for Jerusalem by a "spirit" of the Lord. On the way he was instructed to aid an Ethiopian minister to understand the scriptures. Later Philip baptized him and was then transported via this "spirit":

And when they came out of the water, the spirit of the Lord caught up Philip; and the eunuch saw him no more, and went on his way rejoicing. But Philip was found at Azotas, and passing on he preached the gospel to all the towns till he came to Caesarea. (Acts 8:39-40)

On one hand there seems to be telepathic suggestion to accomplish something, and on the other a physical movement from one location to another. Obviously two different forces were at work. Recall in the earlier abduction cases the telepathic communications used by the aliens to "talk" to those abducted. The "spirit" is similar to these telepathic communications on one hand.

But then there is the physical movement of a person from one place to another, a quite different force. This mysterious thing also happened to Elijah, who rode in the vehicles several times. Ezekiel, too, experienced similar rides, as did Enoch, who *"walked with God, and was not, for God took him,"* (Gen. 5:24) and Jesus. These cases will be more closely examined in a later section in order to understand them in context, where they seem to imply that the patriarchs were flown in the Starmen's vehicles. This God truly had powers which caused Israel to proclaim:

There is none like God O Jeshurun, who rides through the heavens to your help and his majesty through the skies. (Deuteronomy 33:26)

The Mission

By redefining the angels as Starmen, the Lord as their leader, and the ever-present clouds associated with them as their vehicles, other important aspects of this alien presence may be recognized, including their organization, how they carried out their mission and possible reasons for their coming to the Earth.

The organization of the angels is a fact of the Bible. Few authorities will argue with the premise that the Lord is the leader and the angels are the followers and implementors of some great plan. Even the evangelical Billy Graham believes that the angels had a specific organization:

We cannot study the subject of angels in the Bible without becoming aware of ranks among the angelic beings. (126)

While there have been numerous works which provide possible rankings for the angels, including such diverse levels as "principalities" and "cherubim," it has already been proposed that the cherubim were not necessarily ranks, but devices used by the angels. There are two very clear ranks, however, and those are the "angel" and the "archangel." The Bible names only Michael explicitly as an archangel, but the book of Enoch names seven: Uriel, Raphael, Michael, Raguel, Saraqael, Gabriel and Remiel. There is evidence that there are other angels named in both Enoch and other non-canonical writings, but we shall stick with this common list since an exact list of archangels is both difficult to verify and unnecessary to the thesis of this book.

The angels were given specific areas of responsibility. For example the archangel Michael is spoken of in the *Book of Daniel* as the angel in charge of Israel:

At that time shall arise Michael, the great prince who has charge of your people. (Daniel 12:1)

It appears that Michael was the Lord's chief advisor in matters concerning Israel and also in matters of war. Not only did Michael and his subordinates intervene several times for Israel as they fought to acquire a homeland, it was Michael who fought with Satan and cast him to the Earth, according to Revelation 12:7-12. He and his angels also fought against the "Prince of Persia":

The prince of the Kingdom of Persia withstood me twenty-one days; but Michael, one of the chief princes, came to help me, so I left him there with the prince of the Kingdom of Persia...But I will tell you what is inscribed in the Book of Truth; there is none who contends by my side against these except Michael, your prince. (Daniel 10:13, 21)

The archangel Gabriel was most frequently characterized as a messenger. His duties also pertained to Israel, but he is associated with the dissemination of important news to key Bible characters. For example he came to the prophet Daniel more than once to deliver important messages:

While I was speaking in prayer, the man Gabriel, whom I had seen in the vision at the first, came to me in a swift flight at the time of the evening sacrifice. He came and said to me, "O Daniel, I have now come out to give you wisdom and understanding." (Dan. 9:21-22)

Gabriel, plainly stated to be a man, came to the prophet "in a swift flight," in his vehicle, to make Daniel understand something very important. For this he was put into a deep trance, much the way that Betty and Barney Hill were treated during their encounter with alien beings:

As he was speaking to me, I fell into a deep sleep with my face to the ground; but he touched me and set me on my feet. He said, "Behold, I will make known to you what shall be at the latter end of the indignation; for it pertains to the appointed time of the end." (Dan. 8:18-19)

Gabriel also came to Zechariah to announce that his barren wife would conceive and bear a child to be called John the Baptist and to the Virgin Mary to tell her of her conception of Jesus. In the *Book of Enoch,* he is also said to be in charge of the cherubim. (Enoch XX:7)

Of the other archangels, little mention is made of their specific duties. For example Raguel is said to be "one of the holy angels who takes vengeance on the world of the luminaries," or Remiel, who was "one of the holy angels, whom God set over those who rise." Since we have no way of knowing what is meant by those passages, they can simply serve to illustrate that other archangels existed with duties not explicitly related to Israel.

There are many unnamed angels who visit the Israelites to carry out various assignments, such as the destruction of Sodom and Gomorrah, the building of the Ark of Noah, and the delivering of messages. In fact, for a project of this size there would need to be many angels organized into a highly centralized order, and the Bible leaves no doubt about the size and structure of this organization:

With mighty chariotry, twice ten-thousand, thousands upon thousands, the Lord came from Sinai into the holy place. (Psalms 68:17)

Who among the heavenly beings is like the Lord, a god feared in the council of the holy ones, great and terrible above all that are round about him? (Psalms 89:6-8)

The Lord came from Sinai and dawned from Seir upon us; he shown forth from Mount Paran, he came from the ten thousands of the holy ones, with flaming fire at his right hand. (Deuteronomy 33:2)

For he will give his angels charge of you to guard you in all your ways. (Psalms 91:11)

With so many angels going here and there working various projects under the direction of the Lord, it was sometimes impractical or impossible to avoid being seen:

Jacob went on his way and the angels of God met him; and when Jacob saw them he said, "This is God's army!" (Gen. 32:1-2)

There are several reasons to believe that this was a military organization with ranks and authorities. The Bible speaks of the "thrones," "dominions," "principalities" and "powers" of God's organization. Although we do not know what a "principality" or a "dominion" is from what little is given, it is clear that they refer to ranking and structure. Perhaps the titles suggest areas of responsibility for the Starmen, not unlike states, counties, prefectures and zones. There is other evidence throughout the Bible to support the idea of military structure too.

First, whenever the angels carry out their work among the Hebrews, they maintain a low profile. Their work is swift and efficient. They avoid too much exposure, and they come and go as discreetly as possible.

Second, they seem never to originate commands, but carry out the orders of the Lord. They speak with economy of words, and when they do speak, it relates directly to the mission.

Third, they saw war as a means to an end. They carefully trained Israel in military organization, battle tactics, reconnaissance and logistics for the explicit purpose of waging war to secure for themselves a place on the Earth among the other races. They even intervened *directly* in some battles when the tides of battle turned against Israel, as will be shown.

One segment in the Old Testament, beginning with the end of the exodus out of Egypt, is devoted to the acquiring of the "promised land, flowing with milk and honey." But that phrase should read "the land flowing with blood and guts" for the slaughter which occurred in the name of the Lord (and with his help) was by the sword of a driven people carefully trained for war!

This God is my strong refuge...He trains my hands for war...and thy help made me great. (II Samuel 22:33, 35-36)

With a project of such magnitude, and over such a great span of time, how is it that the same God and angels appear to follow Israel from year to year over hundreds and thousands of years? This problem can be resolved by understanding four points: (1) the time spans represented in the Bible, (2) the theory of time dilation in space travel, (3) the Starmen's extended lifespans, and (4) the eventual standardization of the name of God from earlier varied forms.

First let us consider the time intervals represented in the chronology of the Bible. We have seen that a direct lineage was traced from Adam to Noah spanning some 1,600 years in Genesis. Then in Genesis chapter 11:10-32, the genealogy is further traced from Shem, Noah's son, to Abraham through ten "generations," where there appears to have been little or no activity by the Starmen.

After a lapse of several thousand years, there appears to again be a good deal of the Starman activity from the calling of Abraham through Isaac to Jacob. During the time of Joseph, son of Jacob, activity again ceases, and the Israelites come under the bondage of the Egyptians for 430 years. (Ex. 12:40)

Again there is a call to Moses to lead the Israelites out of Egypt to begin the exodus. Throughout the exodus and during the conquest of Canaan, the presence of the Starmen is especially evident. But even the invasion of Canaan became a two-staged operation. At about 1210 B.C., Joshua led the first assault. After this initial thrust, Israel mysteriously halts and remains a loose confederation of tribes for perhaps 200 years until the reign of Saul. From Saul to David, Israel finally becomes fully established in Canaan, their promised land.

Then the presence of the Lord is again conspicuously missing from the reign of Solomon until about the time of Elijah at about 870 B.C., a span of perhaps 100 years. Another obvious lapse comes between the times of Elijah and Isaiah, the prophets, roughly 120 years, with another 100 years between Isaiah and Ezekiel. Finally there is the "seventy weeks of years" prophecy of Daniel 9:24, or 490 years, in which the Lord seems to have left the Israelites on their own until the coming of Jesus. Looking at the whole pattern, we thus have something like this:

Adam to Noah .... 1,600 years active (?)
Shem to Abraham .... 6,000 year lapse (?)
Abraham, Isaac, Jacob .... 100 years active
Joseph to Moses........... 430 year lapse
Moses to Joshua .... 50 years active
Joshua to Saul .... 200 year lapse
Saul to David .... 70 years active
Solomon to Elijah .... 100 year lapse
Elijah .... active
Elijah to Isaiah............ 120 year lapse
Isaiah .... active
Isaiah to Ezekiel........... 100 year lapse
Ezekiel, Daniel .... 50 years active (?)
Daniel to Jesus............ 490 year lapse
Jesus .... active

These are approximations based upon historical and biblical evidence and not hard time periods. But the pattern is important to note for two reasons: First, it illustrates that the Lord was sometimes present and sometimes not. What this means in terms of the Starman Theory is that the Starmen had conquered space so completely so as to come to the Earth to colonize it and they must also have had the ability to travel the great distances of space and return in times which for them were probably not significant.

The idea of traveling the great distances between the stars is theoretically possible according to Einstein's Special Theory of Relativity. According to this theory, no object can travel faster than the speed of light, which is 186,282 miles per second. But the closer that one can get to the speed of light, the slower time will pass for the person traveling at that speed:

> *A most interesting consequence of time slowing down is the following. If an astronaut could travel through space at a really high speed, let us say 99.995 percent of the speed of light, his rate of aging would be slowed by 100 times; so when he returned after the absence that seemed to him, say, 5 years, the earth would meanwhile have aged 500 years, and he could meet his descendants 15 generations down the line!* (127)

This is precisely what may have transpired during those lapses of hundreds of years of Israel's history. There are many stars like the sun in our stellar neighborhood and which are under 15 light years from the Earth. Recall that a light year is the distance light would travel in one year. Traveling at nearly the speed of light would put such stars as Rigil Kent (4.3 light years), Wolf 359 (7.7 light years), and Tau Ceti (11.9 light years) well within reach of the Earth. Also note that many other nearby stars considered to be similar to our sun were identified by Majorie Fish from data supplied by Betty Hill's drawing of a stellar map which she recalled from her abduction by alien beings. Nearly all of the stars she listed fell within 48 light years of the earth, and a third were under 30 light years.

Another aspect, already discussed, that would lend weight to this theory are the reported lifespans of the pre-flood patriarchs, which reached into the 900-year range. If the Hebrews were hybrids of the Starmen, then it would be logical to deduce that the Starmen could live at least as long as Methuselah's 969 years and probably longer. Therefore even if the Starmen could travel through space only at a speed that would slow time by a factor of 10, the trip would still be feasible for them.

This is the logical explanation for why the Lord would sometimes answer the calls of his people and at other times would appear to ignore them. Sometimes the Starmen were simply not present to hear their calls:

And Saul inquired of God, "Shall I go down after the Philistines? Wilt thou give them unto the hand of Israel?" But he did not answer him that day. (I Samuel 14:37)

And the people of Israel inquired of the Lord (for the ark of the covenant of God was there in those days, and Phinehas the son of Eleazar, son of Aaron, ministered before it in those days), saying, "Shall we yet again go out to battle against our brethren the Benjaminites, or shall we cease?" And the Lord said, "Go up; for tomorrow I will give them into your hand." (Judges 20:27-28)

Yet thou hast cast us off and abased us, and hast not gone out with our armies. Thou hast made us turn back on the foe. (Psalms 44:9-10)

Why doest thou stand afar off, O Lord? Why doest thou hide thyself in times of trouble? (Psalms 10:1)

Now the boy Samuel was ministering to the Lord under Eli. And the word of the Lord was rare in those days; there was no frequent vision. (I Samuel 3:1)

The second reason the lapses are significant is that they ultimately caused the mission to fail. Several books of the Old Testament are devoted primarily to the recording of the laws and ordinances which were imposed upon the Israelites by the Starmen in order to form them into a more refined, and manageable culture. But the Starmen, it appears, underestimated the strength of the outside social, cultural and religious pressures on their people.

When the Starmen gave the people rules and regulations, and performed mighty feats to impress upon them the gravity of the mission, they fully expected those laws to be internalized permanently and completely. However, when they departed, a small and vulnerable race was left to fend for itself amongst many other more powerful cultures, and this vacuum resulted in the inevitable mixing of the cultures and perhaps the races as well. Each time the Starmen returned, they found that the Hebrews had corrupted themselves or been invaded, and the Lord expresses no end of frustration over this. Because of the Starmen's miscalculations about the tenacity of the Hebrews, and the Hebrew's failure to adhere to the laws of the Lord, the mission eventually failed, as will be shown.

Finally, if the Starmen did come and go, it is possible that they served tours of duty. If the mission was an on-going project, then the continuity of behavior among the various contingents of Starmen would have led the Hebrews to conclude that they were dealing with the same God throughout their history.

To add weight to this idea, it must be realized that God had been portrayed by perhaps 30 different names in the Bible, from "God of Heaven" to "Yehweh" to "Lord God" to "Lord of Hosts" to "Avarat." Truly it would be foolish to automatically assume that the God of Adam was the same God known at the time of Abraham, Moses or Jesus given the great expanse of time involved and our assumption that the angels were living beings. Indeed there appear in the Bible passages which suggest that there really were several "Lords," and this may even have been alluded to by God himself:

God said to Jacob, "Arise and go up to Bethel, and dwell there; and make an altar to the God who appeared to you when you fled from your brother Esau." (Genesis 35:1)

Behold, the Lord your God has set the land before you; go up, take possession, as the Lord, the God of your fathers has told you; do not fear or be dismayed. (Deut. 1:21)

There can be little doubt that the Bible records a mission of Starmen to establish their hybrid descendents on the earth. Their organization, physical characteristics and temperaments all testify to this theory. But there is still the question of why the Starmen came to this planet.

What could have been the Starmen's purpose in establishing a race of their offspring on earth? Very little is said directly about this in the Bible obviously because the Hebrews were the subjects of the manipulation and not privileged to such information. But there are a few passages which might shed some light on this question:

And I have put my words in thy mouth, and I have covered thee in the shadow of mine hand that I may plant the heavens, and lay the foundations of the earth, and say unto Zion, thou art my people. (Isaiah 51:16)

All the earth is mine, and you shall be a kingdom of priests... (Exodus 19:6)

It appears that they came to "plant" the Earth (and other planets) with their seed. This "seed," or people of Israel, was the sole concern of this God. He gave the Hebrews laws and a system of values that they might proliferate and become a "kingdom of priests" to teach the rest of the peoples of the Earth important moral lessons.

But was this little more than the manifest destiny mentality of the eighteenth century missionaries who traveled to Africa, Asia and the Americas to impart their superior religious morals to the "savages," or would that be oversimplifying the situation? Perhaps the Starmen foresaw our future ventures into space and thus had a greater plan in mind which we have yet to discover.

Methods

There were four principal methods of control used by the Starmen to influence Israel: (1) psychological control, (2) rule by proxy, (3) a military organization, and (4) biological intervention. These were the "tools" employed to carry out the mission with as little direct contact with the masses as possible. The significance of these methods is that they point to a need for separation from the masses for better control and for health reasons, thus strengthening the physical Starman theory.

The first method of control of the ancient Hebrews was through a sophisticated set of *psychological mechanisms* designed to influence the masses into performing in the desired manner. There were four primary mechanisms used to accomplish this: (1) *miracles,* (2) *fear,* (3) *ritual* and (4) *dependence.*

The term "miracles" as used here refers to the great feats performed by the Lord at critical times for the benefit of Israel, such as the parting of the Red Sea, the halting of the flow of the Jordan River, the toppling of the walls of Jericho, the routing of entire armies of enemy soldiers through the Lord's "terrible weapons of indignation," the staying of the sun, and other similar displays of power or deception. These are the kinds of events described earlier which may have been declared miracles by reason of the differences in technology between the Starmen and the Israelites. While better understood and appreciated in the contexts in which they occurred, the significance of miracles at this point derives from the *psychological leverage* they gave to the Starmen for the accomplishment of their mission. Through these awesome displays, the Starmen sought to impress upon the Israelites the power of God. They were explicitly designed to awe the people and to influence by reason of respect for his power:

For the Lord your God dried up the waters of the Jordan for you until you passed over, as the Lord your God did to the Red Sea, which he dried up for us until we passed over, so that all the peoples of the earth may know that the hand of the Lord is mighty; that you may fear the Lord your God forever. (Joshua 4:23-24)

Fear is the second aspect of psychological control used by the Starmen. This "fear factor" has often been brushed aside by Christians as the fear of losing God's blessing or the fear of offending God in some way, but such understanding of fear is clearly inconsistent with that which the Hebrews knew in the Old Testament times. The Starmen's mission required that the Israelites act expediently and consistently in the manner prescribed. There was little time for mistakes, forgetfulness or resistence to the plan, and the Lord showed little patience toward his people when they erred. From the beginning it was stressed that Israel was in for a test, and they were made to understand that they had nothing to fear if they followed the program:

Do not fear; for God has come to prove you, and that the fear of him may be before your eyes that you may not sin. (Exodus 20:20)

To demonstrate his commitment, the Lord made examples of those who were slow to heed his commandments:

While the people of Israel were in the wilderness, they found a man gathering sticks on the Sabbath day. And those who found him gathering sticks brought him to Moses and Aaron, and to all the congregation. They put him in custody, because it had not been plain what should be done to him. And the Lord said to Moses, "The man shall be put to death; all the congregation shall stone him with stones outside the camp." And all the congregation brought him outside the camp, and stoned him to death with stones, as the Lord commanded Moses. (Numbers 15:32-36)

Now an Israelite woman's son ... and a man of Israel quarreled in the camp, and the Israelite woman's son blasphemed the Name and cursed. And they brought him to Moses... And they put him into custody, till the will of the Lord should be declared to them. And the Lord said to Moses, "Bring out of the camp him who cursed; and let all who heard him lay their hands upon his head, and let all the congregation stone him... He who blasphemes the name of the Lord shall be put to death." ...Thus the people of Israel did as the Lord commanded Moses. (Leviticus 24:10-23)

Although such discipline was seen as necessary, God clearly preferred to reward Israel for a job well done. Unfortunately, the people often faltered, and the Starmen were perplexed at Israel's short memory:

"Come now, let us reason together," says the Lord... "If you are willing and obedient, you shall eat the good of the land: but if you refuse and rebel, you shall be destroyed by the sword." (Isaiah 1:18-20)

It is therefore no wonder that all Israel sang:

Serve the lord with fear, with trembling kiss his feet, lest he be angry, and you perish by the way; for his wrath is quickly kindled. (Ps. 2:11)

The thing that becomes crystal clear upon the study of this subject is that the Hebrews had no choice of whether or not to follow the Lord. They were required under penalty of destruction to follow his plan!

Out of heaven he let you hear his voice that he might discipline you. (Deut. 4:36)

And if by this discipline you are not turned to me, but walk contrary to me, then I will walk contrary to you, and I myself will smite you seven-fold for your sins. (Lev. 26:27-28)

No other nation knew this fear. These passages recur consistently throughout the Old Testament, particularly during the exodus and conquest periods, and serve to illustrate not only how fear was employed to influence Israel, but also give some insight into the emotional temperament of God, viz., the

Starmen. This was no less than coercion, and the tribes of Israel were bound to abide by their special status as the Lord's chosen people:

He declared his word to Jacob, his statutes and ordinances to Israel. He has not dealt thus with any other nation; they do not know his ordinances. (Ps. 147:19-20)

The third form of psychological control employed by the Starmen was the *ritual.* Through this mechanism the Starmen accomplished two things: First, the ritual was a way of *psychologically distancing* the Starmen from the masses. It was performed in a manner which tended to place the Starmen far above the masses in status and glory, thus magnifying the Lord's influence over the people.

Second, the ritual *masked direct interaction* between the Hebrew proxy leader and the Starmen from the masses so as to prevent the masses from venerating the forms of the Starmen or their equipment as idols. The less seen of such things, the more omnipresent and omnipotent they would seem.

So important was this ritual that two books of the Bible are devoted primarily to it as well as to the associated rules and ordinances. *Leviticus,* the third book of the Old Testament, is concerned mostly with the instruction of the tribe of Levi, which was chosen by the Lord to assist the high priests. The fifth book, *Deuteronomy* (second law), deals with the instruction of the masses. Minute details are laid out. From the preparation of the offering to rules of association and marriage, to sanitation, to laws and ethical values, it was no less than an attempt at a complete culture synthesis.

The ritual procedure was organized into a hierarchical arrangement with roughly four levels: The Hebrew leader, the high priests, the Levite attendants, and the Hebrew masses. Each successive step from the leader down involved a decreasing level of proximate exposure to the Starmen. The proxy leader had the most intimate working relationship with the Starmen, enjoying knowledge not divulged to any other members of the Hebrew nation. He was the charismatic focus of the people through which the Starmen revealed their word and through whom feedback on various plans and ideas made it back up to them.

The high priests were the elite class. They, along with the proxy leader, had a high degree of understanding of the relationship between the Hebrews and the Starmen and were responsible for administering directly to them when they were present. (Mal. 2:7) The Levite attendants, conversely, probably did not fully understand the relationship although they would have known far more than the masses. They were appointed by the Lord to help the high priests in their duties to the Starmen and acted as intermediaries between the religious ritual of the masses and the service to the Starmen.

The ritual for the worship of the Lord by the masses was largely symbolic. It was designed psychologically to stimulate and reinforce the belief in the Lord's omnipotence by allowing him to remain insulated from the masses, direct contact occurring only through the Hebrew proxy leader and the high priests. Their rules and ordinances were the guidelines. Their ignorance of the true nature of the Lord (as a living being) and their fear of Him were the control mechanisms. This was the two-fold purpose of the ritual in the Intermediate Order.

Despite the grim picture of subjugation painted thus far, it can be said that God's harsh attitude toward his people was balanced with generous concern for their welfare:

"Behold, I am with you and will keep you wherever you go..." (Gen. 28:15)

The Israelites, likewise, truly respected and loved their God, and they reciprocate those emotions with magnified praise throughout the Old Testament:

Blessed be the Lord, my rock, who trains my hands for war, and my fingers for battle; my rock and my fortress, my stronghold and my deliverer, my shield and he in whom I take refuge, who subdues the people under him. (Ps. 144:1-2)

Praise the Lord! Praise the Lord from the heavens, praise him in the heights! Praise him, all his angels, praise him, all his hosts! (Ps. 148:1-2)

Israel, though sometimes reluctant, knew that when the Lord commanded them to carry out a battle or a strategic mission, he was their faithful and powerful ally. With such a great and terrible friend, Israel's small size was no liablity for the Lord was "on call" whenever they required his help up through the time of their establishment in Canaan.

For what great nation is there that has a god so near to it as the Lord our God is to us, whenever we call upon him? (Deuteronomy 4:7)

This confidence in the Lord as an ally enabled the use of the fourth method of psychological control: *dependence* on aid from the Starmen, which could be extended or withdrawn according to Israel's needs and cooperation.

Beyond this the Hebrews were continually made to understand their special position among the races of mankind and their privilege in having the Lord's special attention. Eventually they became totally oriented toward their God, their culture and the land to which they were given claim. Thus it is easy to see that when the Israelites did falter by corrupting themselves in some manner abominable to God and were abandoned or denied God's help, they were totally lost:

Why doest thou stand afar off, O Lord? Why doest thou hide thyself in times of trouble? (Ps. 10:1)

My soul thirsts for God, for the living God. When shall I come and behold the face of God? My tears have been my food day and night, while men say to me continually, "Where is your God?" (Ps. 42:2-3)

O Lord God of Hosts, how long wilt thou be angry with thy people's prayers? ...Restore us, O God of hosts; let thy face shine, that we may be saved! (Ps. 80:4, 7)

* * *

The second method of control of the Hebrews was through *proxy leaders.* Instead of intervening directly to lead and manipulate the people, the Starmen carefully selected and trained leaders from among the tribes of Israel to lead the nation as their direct representatives. This approach had several advantages. First, it allowed the Starmen to maintain their isolation from the masses, as stated earlier, to aid in exploiting the people's ignorance about the mission. This would explain why the Starmen took such pains to camouflage their vehicles in "clouds" or to remain inside their vehicles when they spoke to people. They wanted no form to be worshipped:

Therefore take good heed to yourselves. Since you saw no form on the day that the Lord spoke to you at Horeb out of the midst of the fire, beware lest you act corruptly by making a graven image for yourselves, in the form of any figure, the likeness of male or female... (Deuteronomy 4:15-16)

Without a physical form to worship, there was only the "unseen God" who "looks down from heaven" upon the Hebrews, which greatly magnified his power in their minds, giving him an omnipresent quality.

Second, it allowed the people to be led by the charisma and trust of a familiar figure at a time when the powerful and terrible force of God was seen to be intimidating and alien.

Third, a representative drawn from the people could provide valuable insight into their values, attitudes and reasoning as well as providing other relevant advice when required.

Fourth, a proxy leader gave the Starmen the freedom to come and go and removed some of the risks involved in trying to lead the masses by themselves. The omnipresent quality generated through the lack of a form to worship would also carry the Hebrews through the lapses as the Starmen came and went through time. Even if the Starmen were not present, the feeling by the masses that they might be effectively extended their control.

Finally, leading through proxies removed the liability of being responsible for a mission or a battle which fell short or failed. All responsibility rested with the people, thus protecting Starman credibility in the eyes of the masses. Without proxies, the Starmen would have found their mission greatly complicated, if not impossible.

* * *

The third method of control was through the *military organization* of the Israelites. A hierarchy of control through a chain of command was orderly and efficient. The fourth book of the Bible, *Numbers,* is primarily devoted to this organization of Israel and immediately begins with the Lord's order to take a census:

The Lord spoke to Moses in the wilderness of Sinai, in the tent of meeting, on the first day of the second month, in the second year after they had come out of Egypt, saying, "Take a census of all the congregation of the people of Israel, by families, by father's houses, according to the number of names, every male, head by head; from twenty years old upward, all in

Israel who are able to go forth to war, you and Aaron shall number them, company by company." (Numbers 1:1-4)

Every head of a tribe became a commander with his own standard, and each tribe camped together as a unit. They did not mix. Authority was further divided into Captains over Thousands and Captains over Hundreds (Num. 31:48). The Levites were not included in the census for their work was with the ritual, the tabernacle and the Tent of Meeting.

There was a specific order for encamping and for marching. The tribe of Judah was the first to march. The Levites marched in the middle of the tribes to protect the Tent of Meeting and other religious artifacts:

Then the tent of meeting shall set out, with the camp of the Levites in the midst of the camps; as they encamp, so shall they set out, each in position, standard by standard. (Num. 2:17)

The tribe of Dan formed the rear guard with the other tribes acting as the left and right echelons. As they moved, they sent out reconnaissance units:

The Lord said to Moses, "Send men to spy out the land of Canaan, which I give to the people of Israel." (Num. 13:2)

Israel would soon engage in battles, and thus had to be well trained. The question of why the Israelites spent forty long years "wandering" in the wilderness is thus easily explainable under the Starman Intervention Theory. For the complex training which the Israelites underwent, three things were required: (1) total isolation to reorient the people toward the Lord's mission without distraction, (2) the privacy to learn the new code without interruption from outside forces, (3) and enough time - a generation - to ensure the masses had internalized what the Lord intended. The people needed to learn to depend upon God as their mighty ally as they trained to become a warrior nation, a strike force to claim the land they were promised. This was the great apprenticeship in the Sinai desert.

* * *

The fourth control mechanism employed to influence the Hebrews was *biological intervention.* This involved not only the creation of man from existing naturally evolved homo sapiens, beginning way back at the Eden base, but also the continued maintenance of the chosen lineage through time in order to ensure their establishment on the Earth. This task was complicated by two principal factors: First, since the Starmen chose to remain unseen, the need to intervene periodically and physically increased their exposure. Second, this race of Hebrews kept corrupting themselves, which made it necessary to intervene to maintain the pureness of the stock.

A recurring theme in the Bible which subtly parallels that of the establishment of Israel in the land of Canaan is the "pure vine" metaphor. Recall that the Bible traces a rather detailed genealogy, a recording of the "family tree" of the Hebrews. It is this family tree, or rather a "righteous branch" thereof, which is metaphorically represented to illustrate God's dealing with the chosen race:

"Yet I planted you a choice vine, wholly of pure seed... (Jeremiah 6:2)

"Behold the days are coming," says the Lord, "when I will raise up for David a righteous branch, and he shall reign as king in the land." (Jer. 23:5)

"I am the true vine, and my Father is the vinedresser." (John 15:1)

This vine had righteous branches and corrupt branches, which occasionally had to be "pruned" or even "cut off."

"Every branch of mine that bears no fruit, he takes away, and every branch that does bear fruit he prunes, that it may bear more fruit." (John 15:2)

The metaphor begins in the Garden of Eden with the creation of the hybrid man by the Lord and the first references to the trees of life and knowledge. Later references to Israel and even other races consistently return to this theme:

Thus says the Lord God: "I myself will take a sprig from the lofty top of the cedar, and will set it out; I will break off from the topmost of its young twigs a tender one, and I myself will plant it upon a high and lofty mountain; on the mountain height of Israel I will plant it, that it may bring forth boughs and bear fruit, and it will become a noble cedar..." (Ezekiel 17:22-23)

Israel was holy to the Lord, the first fruits of the harvest, and all who ate of it became guilty. (Jer. 2:3)

But this metaphor extends to the very core of the Bible story if we recall that the Lord's aim was to "plant the heavens and establish the foundations of the earth." In short, the whole Bible, from Adam to Jesus, was oriented toward this theme of the vine which was planted and tended by the Lord, who ceaselessly reminds the Hebrews that they are His people:

"O Israel... you only have I known of all the families of the earth." (Amos 3:2)

The pure vine metaphor stresses the integrity of the Hebrew "family," which includes not only cultural and religious purity, but racial purity as well:

When the Lord your God brings you into the land which you are entering to take possession of it, and clears away many nations before you, the Hittites, the Girgashites, the Amorites, the Canaanites, the Perizzites, the Hivites, and the Jebusites... you shall not make marriages with them... (Deut. 7:1,3)

The patriarchs continually admonished their people not to marry the Canaanites or the Hittites because to do so would corrupt the race.

But the management of this "pure vine" goes further than simply "pruning" or "cutting off." Branches of the vine were made to "rise up" as well, which meant that the Lord in some way ensured that a favored family would predominate to deliver Israel or to benefit it in some way. It is on this raising up of specific branches that we will focus, for the method of ensuring the success of a particular line often meant intervening into the normal process of conception among the chosen.

By studying the births of 18 of the most prominent biblical leaders, from Noah to Jesus, we find that at least *half* were born under special circumstances. For example three generations of wives, from Abraham's wife Sarah to Jacob's wife Leah, were initially barren, yet each gave birth to a child after a visitation by a Starman! Following is a listing of significant births:

	Father	*Mother*	*Child*	*Passage Ref.*
1.	Lamech	Bat-Enosh	Noah	I Enoch CVI:1-19
2.	Abraham	Sarah (barren)	Isaac	Gen. 21:1-3
3.	Isaac	Rebekah (barren)	Jacob	Gen. 25:21
4.	Jacob	Leah (barren)	Ruben	Gen. 29:31
5.	Jacob	Rachael (barren)	Joseph	Gen. 30:20
6.	Manoah	wife (barren)	Samson	Judges 3:2
7.	Elkanah	Hannah (barren)	Samuel	I Sam. 1:5, 20
8.	Joachim	Anna (barren)	Mary	I Mary 2:3
9.	Zechariah	Elizabeth (barren)	John	Luke 1:7
10.	God	Mary (virgin)	Jesus	Luke 1:35

Comparing this list to a list of 18 key biblical figures who existed when Starman activity can be clearly deduced from the scriptures, we can more clearly understand the relationship between these births and the Lord's manipulation of the family tree of Israel:

Eighteen Key Figures in Israel's History

** indicates a documented unique birth*

	Adam - Lamech		**Old Order - prehistorical**
1.	* Noah		
	______ Lapse in starman activity		
2.	Abraham	c. 2166 B.C.	
3.	* Isaac	c. 2066	
4.	* Jacob	c. 2006	**Arising of the Nation**
5.	* Joseph	c. 1915	
	______		Lapse
6.	Moses	c. 1526	**Exodus and Reorientation**
7.	Joshua	c. 1406	
	______		Lapse
8.	* Samuel	c. 1105	
9.	* Samson	c. 1075	**Conquest of Homeland**
10.	Saul	c. 1050	
11.	David	c. 1010	
12.	Solomon	c. 970	
	______		Lapse
13.	Elijah	c. 875	
14.	Isaiah	c. 740	**Divided Kingdom and Exile**
15.	Ezekiel	c. 593	
	______		Lapse
16.	* Mary	c. 15	
17.	* John the Baptist	c. 6	**Preparation for the New Order**
18.	* Jesus	c. 5	

This is not an all-inclusive list of the significant patriarchs. For example, the period lapses shown above cover the period of the Judges and the period of the divided kingdom, when several significant leaders were chronicled, but no Starman activity (nor of God or angels) was indicated from the scriptures. There are other interesting hints of unique types of births, such as the birth of Abraham, which is said in Hebrew legend to have been very similar to that of Jesus, and the birth of Solomon, who was spoken of in the Bible as God's *son* (I Chron. 22:9-10), but we shall concentrate on those specifically mentioned in the Bible or Apocrypha to illustrate the Starman Intervention Theory.

It is interesting to note two points about these births. First, in nearly every case, the women were formerly "barren" until the Lord "opened their wombs" to allow them to conceive. The only exceptions were the births of Noah and Jesus. The other notable point is the fact that in nearly every case either the Lord or an angel visited the afflicted families before the birth of the child. Let us examine some of these cases.

The oldest case we shall inspect deals with the flood patriarch Noah. Recall that when Noah was born to Bat-Enosh, wife of Lamech, he was fearfully strange to his father, so much so that Lamech felt it necessary to go to *his* father Methuselah, who had close dealings with the Lord, to ask him about his son. He said:

> *I have begotten a strange son, diverse from and unlike man, resembling the sons of the god of heaven; and his nature is different, and he is not like us, and his eyes are as the rays of the sun, and his countenance is glorious. And it seems to me that he is not sprung from me but from the angels.* (I Enoch CVI:5-6)

Methuselah is quick to reassure Lamech that Noah is his son and that he need not worry for God would use him to "do a new thing on the earth." How do we reconcile the fact that Lamech was sure that his son "was sprung from the angels" and yet was said to be "his son"? This old apocryphal account from the *Book of Enoch* is perhaps the first account of the rising up of a "righteous branch" from among the Lord's people for their eventual establishment on Earth and can be explained in terms of common medical practice used today all over the world—artificial insemination. Therefore while Methuselah could claim that Noah was indeed the son of Lamech, by his wife, it was equally true that he was "sprung from the angels" through artificial insemination, as evidenced by Lamech's description of the boy.

The next great leader of the Hebrews, chosen by the Lord to begin forming the nation of Israel, was Abram (later Abraham). When Abraham and his wife, Sarah, were advanced in years beyond their normal age for bearing children, three mysterious men came to them. Abraham, who was sitting at the entrance to his tent, immediately recognized them as the Lord and two angels. The men were in a hurry, for they were on their way to destroy Sodom and Gomorrah, but Abraham persuaded them to stop and eat before going on, so they acquiesced and turned aside. While they were there, they also gave Abraham and his wife some startling news:

They said to him, "Where is Sarah, your wife?" And he said, "She is in the tent." The Lord said, "I will surely return to you in the spring, and Sarah your wife will have a son." (Gen. 18:9-10)

Later we learn that *"The Lord visited Sarah as he had said, and the Lord did to Sarah as he had promised. And Sarah conceived and bore Abraham a son..."* (Gen. 21:1) The child was called Isaac.

Isaac was forty years old when he took a wife named Rebekah. But after twenty years of marriage, they began to despair for Rebekah could bear no children; she was barren:

And Isaac prayed to the Lord for his wife, because she was barren; and the Lord granted his prayer, and Rebekah his wife conceived. (Gen.25:21)

Two children were born at that time, Esau and Jacob. Esau took a wife from among the forbidden Canaanites, and Jacob went on to become the favored branch.

When Jacob finally took a wife, named Rachael, it was discovered that she too was barren, which was considered a curse for the ancient Hebrews, for children, especially male children, were important as heirs to the family name and fortune. But the Lord eventually had compassion on Rachael and granted her a child:

Then God remembered Rachael, and God harkened to her and opened her womb. She conceived and bore a son...and called his name Joseph... (Gen. 30:22)

Another interesting example of these aided conceptions occurred to the parents of the mighty Samson:

And there was a certain man of Zorah, of the tribe of Danites, whose name was Manoah; and his wife was barren and had no children. And the angel of the Lord appeared to the woman and said to her, "Behold, you are barren and have no children; but you shall conceive and bear a son. Therefore beware, and drink no wine or strong drink, and eat nothing unclean." (Judges 13:2-4)

The woman immediately went to tell her husband about this:

A man of God came to me, and his countenance was like the countenance of an angel of God, very terrible; I did not ask him whence he was, and he did not tell me his name; but he said to me, "Behold, you shall conceive and bear a son." (Judges 13:6-7)

In a similar fashion Hannah, barren wife of Elkanah, was "visited" by the Lord and bore Samuel, and Elizabeth, wife of Zehariah, bore John the Baptist. But the most famous, and certainly one of the most interesting of these is the virgin birth of Jesus.

An interesting controversy has followed the biblical narration of the birth of Jesus because, despite the fact that Joseph is credited with being the proper descendant from the line of David in the Hebrew family tree, virtually no mention is made in the Bible of Mary's background! According to the well-known story, Mary conceived without the help of her terrestrial husband, which makes Joseph's family lineage curiously irrelevant from the standpoint of substantiating Jesus' background.

But there is an apocryphal book entitled *The Gospel of the Birth of Mary* which reveals that Mary was indeed descended from the line of David. There has been some debate on whether or not this book is authentic; however, it is obvious that the debate is tied to a later theological controversy centering around the legitimacy of the virgin birth of Jesus and not to the authenticity of the gospel itself.

In the primitive ages there was a Gospel extant bearing this name [The Gospel of the Birth of Mary], attributed to St. Matthew, and received as genuine and authentic by several of the ancient Christian sects. It is to be found in the works of Jerome, a Father of the Church, who flourished in the fourth century, from whence the present translation was made. (128)

Faustus, an ancient theologian from the third century A.D. who eventually became the Bishop of Reiz, argued that Jesus did not become the "Son of God" until he was baptized in the river Jordan by John the Baptist around 26 A.D. Evidently, some questioned the idea of the miraculous conception of Jesus even at that early date. The circumstances surrounding the birth of Mary are relevant to the controversy over the virgin birth and, to be properly understood, we must briefly examine the story of the birth of Mary.

According to the *Gospel of the Birth of Mary,* Mary came from the line of David, was born in Nazareth and was educated at the temple in Jerusalem. Joachim, her father, was from Nazareth, and her mother, Anna, was from Bethlehem. Together Joachim and Anna lived for twenty years "chastely in the favour of God, and the esteem of men," but they could have no children. In a now familiar sequence of events, an angel of the Lord appeared to Joachim and Anna to bring them news of the birth of a daughter though Anna was said to have been "barren":

Afterwards the angel appeared to Anna [Joachim's] wife saying: "Fear not, neither think that which you see is a spirit. For I am the angel who hath offered up your prayers and alms before God, and am now sent to you that I may inform you that a daughter will be born unto you, who shall be called Mary." (Mary III:1-2)

Careful instructions are given as to how the child was to be reared, for she was a keystone to an important mission, the future human vessel for the body of Christ. It is made abundantly clear that Mary is fully aware of her special status, if not her mission, because she is in constant contact with the angels of the Lord:

For she every day had the conversations of angels, and every day received visitors from God, which preserved her from all sorts of evil, and caused her to abound with all good things. (Mary V:2)

Finally the time of Mary's duty to the Lord had come, when she was perhaps fourteen years old. At this age she was declared "properly mature" and was sent home to her parents in Galilee. There the angel Gabriel visited her to declare to her the mission, saying *"Hail, O favored one, the Lord is with you!"* (Luke 1:28/Mary VII:3) Mary, who was quite familiar with the visitations by angels, was nevertheless troubled by the angel's words. But the angel came to Mary to tell her that she would conceive "without sin," bring forth a son, and had nothing to fear.

These visitations, the conceptions and births could be dismissed as legends or as "miracles" but for several explicit and dramatic statements made in the Bible and Apocrypha which indicate that the patriarchs (and matriarchs) at least knew of and possibly understood that the Starmen were "magically" causing formerly "barren" women to conceive as a blessing to the family and an assurance of the lineage:

Thus says the Lord who made you, who formed you from the womb... (Isaiah 44:2)

And now the Lord says, who formed me from the womb to be his servant... (Isaiah 49:5)

Now the word of the Lord came to me saying, "Before I formed you in the womb I knew you..." (Jer. 1:5)

What has historically been characterized as God's spiritual "all knowing" manifestation can be easily explained today in terms of the Starman Intervention Theory as evidence of physical beings practicing genetic engineering to produce and manipulate a unique race of people, raising up a "righteous branch" from among existing normally evolved Earth races through such practices as artificial insemination, implanted embryos, test tube pregnancies, etc. This rather strange preoccupation with the biological aspects of the race is an important factor in favor of a physical "god" rather than a spiritual one. In Luke 1:35/Mary VII:19 we read:

The Holy Spirit will come upon you, and the power of the Most High will overshadow you; therefore the child to be born will be called holy, the Son of God.

Given the knowledge of today, we may deduce that the Starmen mildly sedated Mary or hypnotized her while she was being artificially inseminated. The idea of being "formed from the womb" by the Lord becomes clear. The *Book of the Birth of Mary* also speaks of this "miraculous" pattern of births as the angel of the Lord tells Joachim why there are barren women:

And so when he shuts the womb of any person, he does it for this reason, that he may in a more wonderful manner again open it, and that which is born appear not to be the product of lust, but the gift of God. (Mary II:5)

He is told that as the mothers of Abraham, Jacob, Samson and Samuel were barren, so too is the mother of Mary, that she may serve the purposes of the Lord. But this biological intervention was explicitly foretold during the Old Order, during the times of Lamech and Noah, so that there can be no doubt that what took place was the use of "heavenly seed" to manipulate the Hebrews:

And it shall come to pass in those days that elect and holy children will descend from the high heaven, and their seed will become one with the children of men. (Enoch XXXIX:1)

Are we reading accounts of the biological engineering of a race of hybrids on the earth? We have only to accept that the Starmen were physical beings to understand that they physically "opened" and "closed" the wombs of the Hebrew women. The techniques of artificial insemination, conception outside the womb, ovary transplants and genetic engineering are common feats today and show every sign of becoming the boom science of the next decades. Biological manipulation of living things, (plants, animals and man) has already produced better grains with more yield, ensured the survival of animals once expected to become extinct, and made possible the birth of children to formerly childless families. We should therefore expect that an advanced race of beings from some other world, capable of travel to earth, would have this capacity also.

Endnotes

125. Miller, *op. cit.,* p. 154.

126. Billy Graham, *Angels: God's Secret Agents,* Pocket Books, New York, 1977, p. 62.

127. "Relativity," *Merit Student Encyclopedia,* Macmillan Educational Corp., New York, 1978, p. 563.

128. Editors, "Lost Books," *op. cit.,* p. 17.

Exodus 14:14-31

Chapter X
CHRONOLOGY OF THE INTERMEDIATE ORDER

Behold, I send an angel before you to guard you on the way and to bring you to the place which I have prepared. Give heed to him and harken to his voice, do not rebel against him, for he will not pardon your transgressions.

— Exodus 23:20-21

The Covenant

The first of the lineage of Hebrews, beginning with Adam (mankind), was the "seed" by which the ultimate goals of the Starmen would be culminated in a race of priests for the rest of mankind on Earth and be "hosts" to the Starmen. The Hebrews had no other purpose but to serve the Lord. From the garden compound in Eden, where their duties were to "dress it and keep it," (Gen. 2:15) to the conquest of a homeland in Canaan, they were the center of the Starmen's project on Earth.

After the fall from the Lord's trust and eventual corruption of most of the line of Adam, it is decided to intervene and destroy the hybrid Hebrews, and the Starmen (angels) who defiled them, in a great flood. But the "righteous branch" of the Adam stock is preserved in Noah and his few remaining families for their eventual reestablishment in Mesopotamia. After the flood many years pass before the Lord again decides to take an active hand in the affairs of his people.

When the sons of Noah (eponyms for various races) left the Ark after the flood and set out to "multiply and be fruitful," they were fulfilling a commandment of the Lord to increase in numbers for his future use. Once they attained sufficient numbers, the Lord chose leaders from a specific line of these people, the Semites, nurturing their culture and race to ensure they maintained their integrity in order to fulfill the plan for the rest of the world. Abraham, Isaac and Jacob thus became the first proxy leaders of their people - direct representatives of God and implementors of his plan during the formation of the Hebrews into a nation.

Who was this Abraham? Although no secular history of his life has yet been uncovered, many things described during his existence concerned real people and actual events. When the

Bible chronicles Abraham's activities, it becomes immediately apparent that there has been a great jump forward in time - probably 6,000 years or so. Egypt by this time had already existed for more than 1,000 years. In the Mesopotamian valley, the Babylonian empire stretched from Padan-aram in the north to the Persian Gulf in the south, and from the Elam in the east to Mari in the west under its greatest king, Hammurabi, (c. 1700 B.C.). Most scholars date Abraham's existence at the time of the Middle Bronze Age, about 2,000 to 1,700 years B.C. The ancient Sumerian city of Ur was probably his early home though he is reputed to have traveled widely throughout the fertile crescent region of the Middle East. During the course of events, his name was changed by the Lord from Abram to Abraham. He is credited with being the "father of many nations" and the founder of the Hebrew nation in particular. (Gen. 17:5)

The first event of interest to us during the Intermediate Order is the Lord's calling for Abram to depart from Haran in the northern-most reaches of the Babylonian empire to go into the land of Canaan. Canaan lies on the eastern shore of the Mediterranean Sea in what is now Palestine and Israel and formed part of the great trade route between Mesopotamia and Egypt. It was a strategic crossroads between two mighty empires with tremendous potential for agriculture, raising livestock and promoting trade by land and sea. But it was not a hospitable place Abram found when he entered Canaan for the first time with his wife, Sari, his nephew Lot, and their herds of livestock:

About 1900 B.C. Canaan was but thinly populated. Properly speaking it was a No Man's Land. Here and there in the midst of plowed fields a fortified keep could be seen. Neighboring slopes would be planted with vines or with fig trees and date palms. The inhabitants lived in a state of constant readiness. For these widely scattered little townships, like veritable islands, were the objects of daring attacks by desert nomads. Suddenly, and when least expected, these nomads were upon them, with indiscriminate butchery, carrying off their cattle and their crops. Just as suddenly they would disappear again into the vast recesses of the desert plains to the south and the east. (129)

This was the wild country Abram found when he arrived, but it held a central place in the Starmen's plan, which was marked by two promises made to Abram. The first was that the Lord would greatly increase the numbers of his family line:

And the [Lord] brought [Abram] outside and said, "Look toward heaven and number the stars, if you are able to number them." Then he said to him, "So shall your descendants be." (Gen. 15:6)

The second promise the Lord makes to Abram is that his descendants will have a homeland in Canaan:

Then the Lord appeared to Abram and said, "To your descendants I will give this land." (Gen. 12:7)

Abram was seventy-five years old when this came to pass, and to mark this commitment, the Lord made a covenant with Abram as he did with Noah in more ancient times. It was to be a pact of allegiance between the Lord and his people and was symbolized through the act of circumcision:

Every male among you shall be circumcised. You shall be circumcised in the flesh of your foreskins, and it shall be a covenant between me and you. (Gen.17:10)

With these simple acts, the Intermediate Order commenced and the formation of the Hebrew nation began.

* * *

One of the best known stories of the Bible is that of the infamous Sodom and Gomorrah. The destruction of these two cities, which once existed south of the Dead Sea in the fertile Jordan Valley near the plain or "vale of Siddim" (Gen. 14:3), has had a great impact on man's contemplation of the consequences of "godlessness" and "sin." But while the story from which these ideas have sprung has obvious moral implications, there is also a dreadful familiarity about some aspects of the actual destruction.

Late one afternoon two angels came into Sodom. Lot, Abraham's nephew, had by this time moved away from his uncle to this southern city, where he tended his flocks in the abundance of the valley. When he recognized two men entering at the gate to the city as angels, he invited them to come to his house to rest. But the angels had come to Sodom on a grim mission, and refused his offer. Lot, however, was not so easily dissuaded:

But he urged them strongly; so they turned aside to him and entered his house; and he made them a feast, and baked unleavened bread, and they ate. (Gen. 19:3)

Later in the evening, a crowd gathered at the door to Lot's house. They demanded to know who these strangers were. Lot urged them to leave in peace, but the people grew restless and began pushing at the door to get at the aliens.

Sensing imminent danger, the angels pulled Lot inside and secured the door. Suddenly the people outside exploded into a frenzy and tried to force their way into Lot's home. Then, when it seemed that Lot and his visitors would become victims of mass violence, the marauders were struck with "blindness" so that they could only grope helplessly about the doorway. The angels hastily gathered Lot and his family to get them out of the city for it was revealed that they had come to destroy it.

Then the men said to Lot, "Have you anyone else here? Sons-in-law, sons, daughters, or anyone you have in the city, bring them out of the place; for we are about to destroy this place, because the outcry against its people has become great before the Lord, and the Lord has sent us to destroy it." (Gen. 19:12-13)

As the morning approached, Lot still tarried, and so the angels told him just to get his immediate family, who were taken to the edge of the city:

And when they had brought them forth, they said, "Flee for your life; do not look back or stop anywhere in the valley; flee to the hills lest you be consumed...Make haste, escape there, for I can do nothing till you arrive there." (Gen. 19:17, 22)

We are told that the sun had risen when Lot finally arrived in Zoar. Then the Lord rained "brimstone and fire" from heaven on Sodom and Gomorrah. Lot's wife, lagging behind, apparently did not heed the warning not to look back and was turned into a "pillar of salt" by the blast. So completely destroyed was the area that when Abram arose that same morning and looked into the valley from where he had talked to the Lord only the day before, he compared the scene to a smoldering furnace!

This terrible event sounds rather chillingly like the atomic bombing of Hiroshima and Nagasaki near the end of World

War II. It is so well reported that we have little trouble identifying key points to substantiate the physical mission hypothesis: the human-like beings, their taking of food, the sense of urgency to accomplish the mission (a time schedule for God?), and the physical destruction of an entire valley by a deadly explosive force that even killed Lot's wife!

In the entire story, there is but one point that at first appears inexplicable. As the crowd becomes impatient and threatens to break down Lot's door, the people are mysteriously struck with blindness. But this could have been caused by any number of methods known to science today, from a gas to a flash of laser light. Such a mission carried out by modern military standards would not send two men into such a situation without some sort of contingent back-up, such as a silent vehicle hovering high overhead ready to lend assistance should the need arise. Indeed, since the Lord "rained" the destruction on the city from "heaven," the logical answer is that the blinding agent also came from above. Since the blinding occurred after Lot and his visitors finally got the door closed, they did not see what occurred outside. All they saw was the after-effect. While it would be useless to speculate on the endless possibilities for causing mass blindness, we should note that it presents no significant problem by today's standards.

Thus we witness the first "pruning of the vine" by the Starmen and the brutal power at their disposal surgically employed to remove a malignant branch. And yet Lot and his family were carefully extracted. As blood relatives of the still rare Abraham stock, they were an important asset to be preserved for the raising up of descendants promised by God in the formation of a nation of priests.

* * *

To Abraham was born Isaac, who in turn became the father of Jacob. While it is quite apparent that the activities of the Starmen are continuing throughout Isaac's and Jacob's lifetimes, few such spectacular demonstrations of their activities as the destruction of Sodom and Gomorrah are reported during this period. The emphasis continues to be upon events concerning the patriarch's family trees and other relatively mundane matters. But there continues among these accounts, almost as an undercurrent, a series of extraordinary incidents involving mainly Jacob and which seem to pop into the normal chronology at fairly regular intervals to remind us of the alien presence. For example it is during this period, about 1929 B.C., that we encounter Jacob's famous "vision" of the ladder extending from heaven.

While Jacob is traveling from Beer-sheba to Haran to escape his angry brother, he stops to rest for the night somewhere in the wilderness near Bethel. After he had fallen asleep, he was awakened by something so foreign that he thought he surely must be dreaming:

> *And he dreamed that there was a ladder set up on the earth, and the top of it reached to heaven; and behold, the angels of God were ascending and descending on it! And behold, the Lord stood beside him and said, "I am the Lord, the God of Abraham your father and the God of Isaac; the land on which you lie I will give to you and to your descendants... Behold, I am with you and will keep you wherever you go, and will bring you back to this land.; for I will not leave you until I have done that which I have spoken to you."* (Gen. 28:12-13, 15)

Jacob is thus introduced to God and impressed with the weight of the Starmen's commitment to this thing. Jacob is also remembered for the time when he wrestled with the angel (Gen. 32:24-28), as already mentioned, and his witnessing of a multitude of Starmen which he declared to be "God's army." (Gen. 32:1-2)

One of the final acts recorded of the contact between the Lord and Jacob was in the wilderness of Padan-aram. There the Lord appeared to Jacob and changed his name to "Israel," telling him to "be fruitful and multiply," to continue to produce a nation, indeed a company of nations, after which time God would give to his descendants a land of their own.

The primary purpose for this emphasis on the multiplication of the race was to produce enough people to eventually build an army which could then be used to take the "promised land," Canaan, by force. So important was this mission of rapid population expansion that concubines were commonly employed in addition to traditional wives to produce the maximum number of children. It also sheds light on the tremendous grief felt by women who could not bear children for their husbands. The order to "be fruitful and multiply," therefore, doesn't apply today as a commandment of God. It applied only to the early Hebrews and only for the period prior to the occupation of Canaan.

After this final meeting, it appears the Starmen depart for a time:

Then God went up from him in the place where he had spoken with him. And Jacob set up a pillar in the place where he had spoken with him, a pillar of stone. (Gen. 35:13-14)

While other scattered references to the Lord are made, it appears that the *physical* presence is lacking. References to God seem to take on a past-tense perspective as recollections of former acts. Joseph, son of Jacob, carried on after his father to rule the increasing numbers of "Israelites" as they began to call themselves. But it is evident that Joseph had little or no contact as his forefathers had. Since Jacob lived long after Joseph assumed the governorship, however, Joseph knew well what was to come to pass at a future time. The Israelites had also during this time begun to migrate to Egypt, where there was work and the standard of living was good. When Joseph had lived to the ripe old age of 110 years, he summoned his brothers and told them:

I am about to die, but God will visit you, and bring you up out of this land to the land which he swore to Abraham, to Isaac and to Jacob. (Gen. 50:24)

Joseph then died and was buried in the traditional Egyptian manner. His descendants would find that their wait for the fulfillment of God's promise would extend for more than four hundred years and that their status in the land of Egypt would change dramatically during that time.

The Exodus

The Hebrews in Egypt were from the beginning considered merely second-class citizens. But as time passed, they found themselves coming under increasing subjugation. About the time of the Exodus, there came a new ruler over Egypt, the Pharaoh Rames II (1301-1234 B.C.), whose vanity was matched only by his appetite for building great monuments. The Hebrews, who had grown quite numerous by this time, thus became a prime source of cheap labor for Pharaoh. Of course the circumstances are substantially more complex, but this is the basic state into which Israel had fallen when the Lord again turns his attention to his people.

Most people familiar with the story of the exodus will recall the account of a young Egyptian princess who discovered a Hebrew infant in the bulrushes and reared him as her own child. He became known as Moses. Perhaps no other biblical personality was so pivotal to the race of Hebrews except for Jesus. So influential was he that Michael H. Hart places Moses sixteenth in his ranking of the 100 most influential people in world history! And yet so little is known of his life except what is provided in the Bible. Keller cites some known facts:

Moses was a Hebrew who was born in Egypt, brought up by Egyptians, and bore a typical name. Moses is the name "Maose," which simply means "boy-son." A number of pharaohs are called Ahmos, Amasis, Thutmos. (130)

Brought up in such a lavish environment, he would have had the benefit of schooling and training virtually unknown to his Hebrew kin. But we learn that Moses eventually became aware of his kinship to the Hebrews, and he found himself increasingly torn between the life of his adoptive hosts and the wretched lives of his people. When he was out touring the construction one day, he was overcome with fury at the cruelty of the Egyptian taskmasters and, seeing the opportunity, killed one and hid his body in the sand. Then he became fearful of being discovered and fled into the wilderness, to a place called Midian, where he eventually took a wife and became a shepherd.

This is the condition the Starmen found when they returned. It was now necessary to retrieve their people from the Egyptian slavery. And what better choice could they have made for a proxy leader to guide their people out of bondage than this educated Hebrew, Moses?

Now Moses was keeping the flock of his father-in-law, Jethro the priest of Midian; and he led his flock to the west side of the wilderness and came to Horeb, the mountain of God. And the angel of the Lord appeared to him in a flame of fire out of the midst of a bush; and he looked, and lo, the bush was burning, yet it was not consumed. (Ex. 3:1-2)

Moses decided to investigate this strange phenomenon. To his astonishment, as he neared the apparition, a voice called out, *"Moses! Moses!"* In bewilderment he answered, *"Here am I,"* and God called to him, *"Do not come near; put off your shoes, for the place on which you are standing is holy ground."*

(Ex. 3:5) Moses hid his face in fear. What had he stumbled onto?

Although this incident is traditionally thought of in spiritual terms, a physical analogy reveals that there is plausible evidence that Moses had actually come upon a vehicle of the Starmen. First, the reference to a bush that is burning but is not consumed would also be the effect produced by a glowing cloud or steam-shrouded spaceship from which man could speak. If seen from behind a bush or shrub, the effect is obvious. The bush appears to burn but is not consumed.

The "burning" object that is not consumed is a characterization for God that is fairly regularly repeated in the Old Testament.

For the Lord your God is a devouring fire, a jealous God. (Deut. 4:24)

This mysterious "fire" is often confused with God himself but, as we already know, this was merely testament to the prevailing ignorance. When Moses first saw the "burning bush" apparition, he was confused because he had no other experience from which to draw, but later passages indicate that at least part of the mystery was resolved in that it became apparent that the Lord and the "fire" were separate:

Did any people ever hear the voice of a god speaking out of the midst of the fire, as you have heard, and still live? (Deut. 4:33)

On the day that the tabernacle was set up, the cloud covered the tabernacle, the tent of testimony; and at evening it was over the tabernacle like the appearance of fire until the morning. (Num. 9:15)

And in the morning watch the Lord in the pillar of fire and pillar of cloud looked down upon the host of the Egyptians... (Ex. 14:24)

Sometimes the Lord is properly mentioned in association with his fire which clearly indicates a separate, physical fire and not the figurative "fire" of the burning bush:

Our God comes, he does not keep silence, before him is a devouring fire... (Ps. 50:3)

A second observation to point out is that surrounding the "angel" is a "holiness" which Moses is warned to stay clear of and to protect himself by ensuring he is properly grounded by removing his shoes. It can be demonstrated that the term "holy" has two meanings. Today it is often used to refer to a condition of ritual sacredness. But in ancient times, it sometimes had no other meaning than a dangerous condition associated with the presence of the Lord from which the people had to be protected:

And when Aaron and his sons have finished covering the sanctuary, as the camp sets out, after that the sons of Kohath shall come to carry these, but they must not touch the holy things, lest they die. (Num. 4:15)

A third observation is that the Lord spoke *from within* the "angel," so that the angel in this account is reduced to a separate entity, supporting the notion that it was really a vehicle which gave the appearance of a burning fire to Moses and produced a dangerous force field around it. There are other events in the Bible which could be compared to this one, such as the bright light described as an "angel of the Lord" which descended at the tomb of Jesus on the day of his ascension into heaven (Matt. 28:1-3) and from which a voice also speaks to two women in the area. And there are other passages which suggest that some Hebrews may have understood that such light or "fire" was only a covering:

Bless the Lord...who coverest thyself with light as with a garment... (Ps. 104:1-2)

Was this an alien vehicle? Taken by itself the burning bush incident means little, but as we continue to analyze these reports, a certain continuity will emerge.

Moses is thus introduced to God, viz., the Starmen. He is also at that time given a mission; he is instructed to go back to Egypt as an emissary of the Lord to secure the release of his people. He was told to perform certain miracles before the Pharaoh in order to impress upon him the power of the Hebrew God and the authority from whence the order originated. For the time being, we cannot be certain whether Moses had yet seen the Starmen in person, but soon he would meet them face to face.

With his brother Aaron as his eloquent spokesman, Moses journeyed to Egypt and, after working the miracles specified by the Lord, finally forced Pharaoh to concede to the terrible wonders of the Hebrew God and let the Hebrews go.

After 430 years in Egypt, the people hastily gathered themselves together. Then, perhaps a million or more in number, they departed with their cattle and what few possessions they could carry toward their destiny. We are thus struck with this incredible image of an entire race of people being led by Moses out into the desert and to their first regrouping at Succoth, where they would learn of a God they had never known! (Ex. 3:13)

As the people begin to move away from Egypt, we realize that what is taking place on the ground is only half of the picture, for it is a "night of watching" by the Lord. From his great vehicle, he surveyed the proceedings and led the way.

And they moved from Succoth and encamped at Etham on the edge of the wilderness. And the Lord went before them by day in a pillar of cloud to lead them along the way, and by night in a pillar of fire to give them light, that they might travel by day and by night; the pillar of cloud by day and pillar of fire by night did not depart from before the people. (Ex. 13:21-22)

Recall from an earlier chapter on UFOs that the "pillar" type UFO is not uncommon in contemporary reports. Indeed the description of a "pillar of fire" quite appropriately characterizes the UFOs reported from France in 1954 and from Massachusetts in 1966.

When Moses first encounters the "burning bush" on Mt. Horeb, he may have witnessed a smaller scout or shuttle vehicle such as those reported in the same cases cited above as entering and leaving the "mother ship." It is interesting to note, too, that the biblical writer seems to have understood that the Lord led the way *in* the pillar and was not the pillar itself.

Having second thoughts about their defeat, the proud Egyptians subsequently decide to disallow their massive pool of cheap labor simply to walk out on them. We are told that elite Egyptian horsemen and war charioteers pursued the Israelites to the shores of the Red Sea (or the Reed Sea). Coming upon the sight of the pillar, however, the warriors became wary and decided to keep watch through the night. Then, at daybreak, they mistakenly prepared to attack.

The people of Israel began to fear for their lives at this point. Trapped between the Egyptian army and a great body of water, the prospects seemed bleak. Although the method is not specifically stated, Moses is in communication with the Starmen, and he is given instructions to calm the people and to watch the work of the Lord on their behalf:

Then Moses stretched out his hand over the sea; and the Lord drove the sea back by a strong wind all night and made the sea a dry land, and the waters were divided. And the people went into the midst of the sea on dry ground, the waters being a wall to them on their right and on their left. The Egyptians pursued, and went in after them into the midst of the sea, all Pharaoh's horses, his chariots and his horsemen. And in the morning watch the Lord in the pillar of fire and pillar of cloud looked down upon the host of the Egyptians, and discomfited the host of the Egyptians, clogging (or binding) their chariot wheels so that they drove heavily; (Ex. 14:21-25)

With the Hebrews safely across, and the Egyptians bogged down in the middle of the divided sea, the Lord released the force, sending the walls of water crashing down upon them and drowning them all.

At least two forces were at work here: The first drove back the waters and was at least accompanied by a great wind. The second force "discomfited" the Egyptians, causing their chariot wheels to bind when the Lord "looked down" upon them. It makes no difference that we cannot explain the forces which were employed - we still cannot explain how UFOs fly either. It is, however, important to recognize the phenomenal physical display orchestrated for the Hebrews.

This event raises two obvious questions: First, if God was not a physical being, but a spiritual one, why did he choose this physical solution as opposed to simply changing the Egyptian's minds or some other intangible strategy? Answer: This god was a physical being. Second, what purpose did this grandiose display serve for the Starmen? This question is neatly answered by two short passages immediately following the record of the parting of the sea:

And Israel saw the great work which the Lord did against the Egyptians, and the people feared the Lord; and they believed in the Lord and in his servant Moses. (Ex. 14:31)

The Lord is a man of war, the Lord is his name. (Ex. 15:3)

Thus the Starman Lord used this highly visible, physical means to demonstrate to the people His power. It had the desired effect: the people were awed. He was their physical strength in their time of need; he was their mighty "man of war."

After their baptism at the Red Sea, the next significant event for the Israelites was to meet their God in all his glory and to receive the Ten Commandments. This was all carefully arranged by explaining to Moses what must be done prior to the meeting:

...for on the third day the Lord will come down upon Mount Sinai in the sight of all the people. And you shall set bounds for the people round about, saying, "Take heed that you do not go up into the mountain or touch the border of it; whoever touches the mountain shall be put to death. No hand shall touch him, but he shall be stoned or shot; whether beast or man, he shall not live. When the trumpet sounds a long blast, they shall come up to the mountain." (Ex. 19:11-13)

It should by this time be obvious that something about the Starmen's environment is unsafe for man or beast, something so toxic that should any living thing come near it at the wrong time, it would have to be destroyed. The object of concern was the alien vehicle:

On the morning of the third day there came thunders and lightnings, and a thick cloud upon the mountain; and a very loud trumpet blast, so that all the people who were in the camp trembled. Then Moses brought the people out of the camp to meet God; and they took their stand at the foot of the mountain. And Mount Sinai was wrapped in smoke because the Lord descended upon it in fire; and the smoke of it went up like the smoke of a kiln and the whole mountain quaked greatly. And as the sound of the trumpet grew louder and louder, Moses spoke and God answered him in thunder and the Lord came down on Mount Sinai, to the top of the mountain, and Moses went up. And the Lord said to Moses, "Go down and warn the people, lest they break through to the Lord to gaze and many of them perish. And also let the priests who come near to the Lord consecrate themselves, lest the Lord break out upon them." Ex. 19:16-22)

This outstanding example of biblical recording gives us the entire sequence of events: the sounds, the physical effects and the precautions taken, all presented matter-of-factly as if it were a news report. We need little imagination to recognize the landing of a vehicle with the Starmen taking great care to see that no one comes near enough to be hurt by the rocket blasts or by any unseen forces in the vehicle's environment which might "break out" upon them.

But this introduction was just the beginning for there was much work to be done to turn this unruly mass of people into an organized congregation. There were many things which they would have to learn and many old ways which they would have to abandon.

Indeed this shock-treatment was designed to break the people's allegiance to any of the useless pagan gods to which they may have previously paid homage as well as to give them the prospect of a fearful ally on whom they could depend. If these people were to be turned into priests, they would have to undergo an intensive period of training for as long as a generation. They would need laws, values, regulations and knowledge on how to care for themselves and their god. They would eventually need a place of their own, a piece of territory which they could own and defend. To prepare for this would require a strong sense of national identity, good organization and a powerful ally. These conditions would be met through the forty years of "wandering" which the Israelites would soon undergo in the privacy of the Sinai wilderness.

Besides receiving the general rules on the first day, the people were also made to understand their position in the new system of things: they were owned by the Lord; they were to follow; Moses was their leader, and the Lord was his authority. For what was about to take place, there was no room for undue delays or questioning:

> *Behold, I send an angel before you to guard you on the way and to bring you to the place which I have prepared. Give heed to him and harken to his voice, do not rebel against him, for he will not pardon your transgression...* (Ex. 23:20-21)

We also begin to realize that Moses had an *intimate* relationship with the Starmen, speaking face to face with them and having much understanding about the nature of their operation:

> *Moses alone shall come near to the Lord; but the others shall not come near...* (Ex. 24:2)

> *Now Moses used to take the tent and pitch it outside the camp, far off from the camp, and he called it the tent of meeting... When Moses entered the tent, the pillar of cloud*

would descend and stand at the door of the tent, and the Lord would speak with Moses... Thus the Lord used to speak to Moses face to face, as a man speaks to his friend. (Ex. 33:7-11)

Leaving no doubt that Moses spoke to the Lord face to face, we read that the Lord once rebuked Miriam and Aaron, two of the high priests in the camps of Israel, because they had questioned the supreme authority of Moses to declare what was right and wrong among the people, saying:

Hear my words; if there is a prophet among you, I the Lord make myself known to him in a vision; I speak to him in a dream. Not so with my servant Moses; he is entrusted with all my house. With him I speak mouth to mouth, clearly, and not in dark speech; and he beholds the form of the Lord. (Num. 12:6-8)

The Ark

Historically, the Ark of the Covenant must have held a rather curious position in the minds of men, that golden receptacle with mysterious powers which served both as a container for the tablets inscribed with the Ten Commandments and as a device for communicating with God. But why communicate through *anything* to speak to a spiritual God? The Ark loses its cloak of mystery, however, if we accept that God was a group of Starmen and translate passages concerning the Ark literally. The Ark was a powerful tool of the Starmen and played an important role in the implementation of their plan for Israel.

First we must understand that the receipt of the Ark marks a second important phase in the organization of the people. After being brought out of Egypt into the wilderness, they were introduced to God. Following the introduction, they began an initial period of basic instruction, receiving rules, laws and concepts to orient the masses. In preparation for this second phase, Moses is called back to the mountain:

Then Moses went up on the mountain, and the cloud covered the mountain. The glory of the Lord settled on Mount Sinai, and the cloud covered it six days; and on the seventh day he called Moses out of the midst of the cloud. Now the appearance of the Lord was like a devouring fire on the top of the mountain in the sight of the people of Israel. And Moses entered the cloud and went up on the mountain. And Moses was on the mountain for forty days and forty nights. (Ex. 24:15-18)

In a confirmation of our hypothesis of the nature of the burning bush incident, we learn that the *appearance* of the "glory" or corona surrounding the vehicle which contained the Lord was *like* a devouring fire.

It appears that Moses was receiving not cursory instructions, but very detailed, formal instruction for the mission which lay ahead. He was given laws, rules and values to impart to the people for their health and for the service to their Lord. It was to be no less than a complete cultural, social and psychological reorientation.

Detailed instructions were also given at that time for the construction of the Ark of the Covenant and its portable sanctuary:

They shall make an ark of acacia wood; two cubits and a half shall be its length, a cubit and a half its breadth, and a cubit and a half its height. And you shall overlay it with pure gold, within and without shall you overlay it...then you shall make a mercy seat (a cover) of pure gold...and you shall make two cherubim of gold; of hammered work shall you make them, on the two ends of the mercy seat...and you shall put the mercy seat on the top of the ark...there I will meet with you, and from above the mercy seat from between the two cherubim that are upon the ark of the testimony, I will speak with you of all that I will give you in commandment for the people of Israel. (Ex. 25:10-22)

A subsequent passage reveals that this was only a basic description of the device for Moses was also given a blueprint:

And see that you make them after the pattern for them which is being shown to you on the mountain. (Ex. 25:40)

There can be no question that the ark was a communications device. The biblical passage tells us this directly: God will not be there. He will speak through the Ark from between the cherubim. Logically it follows that the human Starmen had their sanctuary or base on Mount Sinai. By placing a communications device in the midst of the Israelites, they could monitor them on their slow marches and during encampment without the dangerous and tedious process of close contact on a continual basis.

There are several reasons for believing this was so. First, the remarkable precautions taken by the Starmen to secure a perimeter around their vehicle on the mountain. Their constant warnings of danger indicate that they might not have been as accessible to the people as they wanted to be. Perhaps they had little control over the forces associated with the vehicle, the force that might "break out" on living things.

Second, the constant movement of Israel through the wilderness, when the Starmen could not always be close at hand, necessitated some mode of communication between themselves and the leadership on the ground. Third, the Starmen's ability to remain "invisible" to the masses by eliminating the need for their physical presence would, as already mentioned, have effectively magnified their psychological control over Israel. Fourth, the precise instructions given for the construction of the Ark and its tabernacle themselves argue for the probability that the Ark was more than a symbolic device. Finally, the best evidence that the Ark was a communications device was that Moses spoke to the Lord through it:

> *And when Moses went into the tent of meeting to speak with the Lord, he heard the voice speaking to him from above the mercy seat that was upon the ark of the testimony, from between the two cherubim; and it spoke to him.* (Num. 7:89)

The Ark thus filled the bill for the Starmen perfectly. It was an ingenious device with a split-level function: For the masses it served as a religious receptacle and the centerpiece for their ritual, and for the Starmen it was a durable communications device which allowed them to remain in their sanctuary on the mountain and yet maintain constant contact with their people through Moses.

To house the Ark, an elaborate tabernacle was also constructed. It was a portable structure which sheltered the Ark while the tribes were at rest and was attended to exclusively by the tribe of Levi. It had two chambers, one for the preparation of ritual offerings and another chamber called the "Holy of Holies" or "Most Holy Place" in which the Ark was kept. Around the perimeter of the sanctuary was a fence to provide limited access to the area. Together the three elements - the ark, the tabernacle and the Levites - formed the ritualistic foundation which served the dual function of reeducating the masses and administering to the needs of the Starmen.

For a time the great pillar UFO would lead the way for Israel continually. Eventually it would be found to come and go, apparently as Moses and the Levites learned to use the Ark. As the physical presence of the Starmen diminishes, the focus of Israel's attention is redirected to the ritual associated with the Ark, the tabernacle and the Levite priests. We can thus begin to form a mental picture of the emerging organization from the following passage:

On the day that the tabernacle was set up, the cloud covered the tabernacle, the tent of the testimony; and at evening, it was over the tabernacle like the appearance of fire until morning. So it was continually; the cloud covered it by day, and the appearance of fire by night. And whenever the cloud was taken up from over the tent, after that the people set out; and in the place where the cloud settled down, the people of Israel encamped. (Num. 9:15-18)

Soon many new rules were imposed upon the people - subjects ranging from the proper ritual offerings to marriage. The strictest rules of cleanliness were drilled into them. The priests of the tribes became medical laymen and were taught to diagnose and separate the diseased from the healthy (all diseased people, and women in menstruation, were banished from the main body):

Thus you shall keep the people of Israel separate from their uncleanness, lest they die in their uncleanness by defiling my tabernacle that is in their midst. (Lev. 15:31)

It may be that the Starmen feared for their own health and thus ordered the purge of the unclean from the camps. While the Starmen rarely dealt directly with the commoners at this stage, they appear to have had contact with the priests and Moses and could have feared contamination through them or through their offerings.

Remarkable details were observed concerning the proper foods which could be consumed and offered to the Lord. The "offerings" were quite interesting in that the preparations were so specifically prescribed that they even included the proper seasonings:

No cereal offering which you bring to the Lord shall be made with leaven; for you shall burn no leaven nor any honey as an offering by fire to the Lord. As an offering of the first fruits you may bring them to the Lord, but they shall not be offered on the altar for a pleasing odor. You shall season all your cereal offerings with salt. (Lev. 2:11-13)

Passages such as this fill many pages of this section of the Old Testament. It has already been suggested that the Starmen were human and ate just as does man. Did this ritual have a practical counterpart as service to the Starmen? The specificity in prescribing the selection, preparation and cleanliness of the various foods, as well as the diversity of the types of food offered to the Lord, takes on a new significance if we recall the oft repeated designation of Israel's god as the "Lord of Hosts." (Is. 47:4) Israel's duty was to host the Lord on his visits to Earth, to serve him: *"You shall fear the Lord your God; you shall serve him, and swear by his name."* (Deut. 6:13)

And so the people commenced their painstaking and tedious process of reeducation to conform to the new order of things. Mistakes were made along the way, some of which were accidents, others of which were miscalculations of the tolerance of the Lord. Several mistakes were paid for with the lives of the Israelites. Consider the following case concerning the death of two priests as a result of a misunderstood ritual process:

Now Nadab and Abihu, the sons of Aaron, each took his censer and put fire in it and laid incense on it and offered unholy fire before the Lord such that he had not commanded them. And fire came forth from the presence of the Lord and devoured them, and they died before the Lord. (Lev. 10:1-2)

Was it the vengeance of the Lord? Probably not. The biblical narrator seems to be reporting the accident as the result of an impersonal force associated with the *presence* of the Lord, as stated earlier, and such forces may have taken their toll simply because specific instructions were not properly followed.

Another example quite different in nature concerned the man mentioned earlier who was part Egyptian, part Israelite. It appears that his Egyptian influence diluted his respect for the Hebrew god for in a dispute with another Israelite, he blasphemed the name of the Lord. The people then seized the man and put him into custody until the will of the Lord was made known concerning him:

And the Lord said to Moses, "Bring out of the camp him who cursed; and let all who heard him lay their hands upon his head, and let all the congregation stone him... He who blasphemes the name of the Lord shall be put to death." (Lev. 24:13-16)

Recall also that a similiar fate befell a man gathering sticks on the Sabbath day (Num. 15:32-36), the day set aside as a day of rest by the Lord. Gradually, however, Israel began to assume its new identity and to accept the new way of doing things.

The Conquest

Once the Starmen perceived that the people of Israel were sufficiently reoriented, they set about to prepare them to go forth into the land which the Lord had promised them for so long. To claim this land, they would have to battle other established peoples and overthrow them. They therefore had to become proficient warriors. This third phase of their desert apprenticeship began with a census of the people approximately one year after their departure from Egypt. Said the Lord:

Take a census of all the congregation of the people of Israel, by families, by fathers' houses according to the number of names, every male, head by head; from twenty years old and upward, all in Israel who are able to go forth to war, you and Aaron shall number them, company by company. (Num. 1:2-3)

The census revealed a population of about 600,000 men, not including the Levites, who had charge of the Ark and the tabernacle. As mentioned earlier, the army was to be made up of 12 tribes, and each was given a specific order and position for marches at this time. The tribe of Judah was to set out first, followed by the Levites, who were flanked by tribes on either side to protect the Ark, with the tribe of Dan forming the rear guard. Behind them followed the stragglers, the crippled, the sick and other assorted refugees who attached themselves to the march.

Eventually the Israelites were forged into a finely tuned military machine. They mastered the tactics, reconnaissance and logistics of war, and had become well disciplined through their wilderness regimen. Their service to their paternal God, and His reciprocal attention to His nation, had by this time lit a fire in the spirit of the people as they began to understand their destiny. They were no longer an unruly band or second-class sojourners, but an army nation with a national pride and a powerful ally. Finally the time arrived for them to enter into their first battle:

The Lord said to Moses, "Avenge the people of Israel on the Midianites; afterward you shall be gathered to your people." And Moses said to the people, "Arm men from among you for the war, that they may go against Midian to execute the Lord's vengeance on Midian." (Num. 31:1-3)

Israel fared well that day, laying waste to Midian, killing all men, women and male children. The booty included precious metals, livestock and young girls. The Lord approved, but there was much to be done. This was only a trial run. A mere 12,000 men went out against Midian to claim this first victory. Soon they would go up against forces much more powerful than themselves. But how could they contend with armies which were much stronger? The answer, amazingly clear in the Books of Deutoronomy and Chronicles, is that the Starmen themselves fought directly for Israel in such cases.

In the first example, a reconnaissance team was dispatched to spy out the land of Canaan. But when the contingent returned, they brought grim news:

The people are greater and taller than we; the cities are fortified up to heaven. (Deut. 1:28)

The Israelites were understandably afraid to go up against a mightier army for they knew that they could not prevail against the superior weapons and manpower of such adversaries, and they were still inexperienced in the art of war. But Moses assured them that they would have a tremendous advantage:

Do not be afraid of them. The Lord your God, who goes before you will himself fight for you, just as he did in Egypt before your eyes. (Deut. 1:30)

Was this some sort of euphemism or allegory for the "spirit" of God enabling the warriors of Israel to achieve victory? No! Other passages confirm that physical intervention actually occurred. A graphic example of this is given during the time of Israel's King Jehoshaphat:

Thus says the Lord to you, "Fear not, and be not dismayed at this great multitude; for the battle is not yours but God's. Tomorrow go down against them; behold, they will come up by the ascent of Ziz; you will find them at the end of the valley, east of the wilderness of Jeruel. You will not need to fight in this battle; take your position, stand still, and see the victory of the Lord on your behalf, O Judah and Jerusalem." Fear not, and be not dismayed; tomorrow go out against them, and the Lord will be with you. (II Chron. 20:15-17)

Other times the people are told not to try to wage war, for the Starmen are not present or have withdrawn their aid:

And the Lord said to [Moses], "Say to them, Do not go up or fight, for I am not in the midst of you; lest you be defeated before your enemies." (Deut. 1:42)

But a man of God came to [Amaziah] and said, "O king, do not let the army of Israel go with you, for the Lord is not with Israel, with all the Ephraimites. But if you suppose that in this way you will be strong for war, God will cast you down before your enemy; for God has the power to help or to cast down." (II Chron. 25:7-8)

And so it was: sometimes the Lord would be present; sometimes he would not be. Sometimes he would aid Israel; other times not. It may be noted that despite Moses' assurances to his commanders of the Lord's allegiance, they still doubted his ability to help them. Because of Moses' inability to rally the people behind the Lord's plan, the entrance into the promised land had to be delayed. Moses by this time had become quite advanced in years and, because of this delay, he never saw the conquest of the promised land.

Moses was a hundred and twenty years old when he died; his eye was not dim, nor his natural force abated. (Deut. 34:7)

An interesting mystery is to be found in the story of the death of Moses, for in the Bible we are told that he died and was buried *by the Lord* in "the valley in the land of Moab" and that "no man knows the place of his burial to this day." (Deut. 34:6)

According to Drake (131), the Talmud claims that Joshua searched everywhere for his teacher and friend but failed to find him. Moreover Josephus, the ancient Jewish historian, implies that Moses was not buried at all, but was taken into heaven *in a cloud.* In his *Antiques of the Jews,* he states that *"a cloud stood over him on the sudden, and he disappeared in a certain valley."* (132) Such a statement is quite significant for the Starman

Intervention Theory because, if he was taken into "heaven" by the Starmen, at least two important passages related to Moses would be clarified. The first deals with how Moses supposedly died.

Though old by our standards, the fact that he didn't act old to the chronicler is important because he actually may not have died at that time. For one thing, the Talmud states that his intimate friend Joshua looked everywhere for Moses but could not find him - a secret burial? Why? Deuteronomy states he was buried by the Lord, and no one knows just where. But if, as stated by Josephus, Moses did not die, but disappeared suddenly when a "cloud" stood over him, perhaps he was taken into "heaven" *alive* and not buried at all.

An intriguing corroboration of this idea is to be found in the story of Jesus' transfiguration, where, to our amazement, a "bright cloud" overshadows Jesus and his companions Peter, James and John on a high mountain. There they witness Jesus talking with none other than Moses and Elijah! Recall that Elijah was also taken *alive* into heaven. (II Kings 2:11) This also ties in with the resurrection of Jesus, when the supposedly dead Christ was taken alive into "heaven" in a "cloud" and is seen a short time later, by Paul and others, *alive.*

Was Moses' death actually an orchestration by the Starmen to remove Moses after his failure to rally Israel's army in the face of the great battles ahead? Was Moses "retired" to allow for the timely succession of his more able lieutenant, Joshua? We know that other proxy leaders were taken into heaven when their honorable service was completed on Earth, such as Enoch, the righteous sage who "walked with God and was not, for God took him." (Gen. 5:24)

In any case Joshua, like Moses, dealt closely with the Starmen and was groomed early on by his predecessor to assume the proxy duties. He was well qualified to lead Israel during the critical period ahead. As a commander from the tribe of Ephraim, he won a brilliant victory against the Amalekites at Sinai (Exodus 17) and was one of the tribal members sent to reconnoiter Canaan. He was also placed in charge of the Tent of Meeting prior to Moses' disappearance, a key position for a successor.

Joshua had to be resolute. He could have neither a faint heart nor a faltering faith in the Lord during the conquest, and he had to be charismatic enough to inspire Israel to follow him into battle. He is therefore strictly admonished not to let the people's confidence waver in the sight of great adversaries, as had occurred under Moses. They were assured of victory so long as they maintained an unshakable faith in their celestial ally:

Moses my servant is dead; now therefore arise and go over this Jordan, you and all this people, into the land which I am giving them, to the people of Israel.

No man shall be able to stand before you all the days of your life; as I was with Moses, so I will be with you; I will not fail or forsake you.

Have I not commanded you? Be strong and of good courage; be not frightened, neither be dismayed; for the Lord your God is with you wherever you go. (Josh. 1:2, 5, 9)

With this charge the warrior Joshua took command and prepared Israel for the crossing of the Jordan River, the traditional beginning of the conquest of Canaan. But this was to be no ordinary river crossing for they would pass over, as at the Red Sea, on dry land! Joshua instructs them:

When you see the Ark of the Covenant of the Lord your God being carried by the Levitical priests, then you shall set out from your place and follow it, that you shall know the way to go, for you have not passed this way before. Yet there shall be a distance between you and it, a distance of about two thousand cubits; do not come near it. (Josh. 3:3-4)

On the third day, led by the Levites, Israel approached the river. Even though it was the flood season, and the waters of Jordan were overflowing, the great flow was halted as if an invisible dam were suddenly erected:

...and when those who bore the ark had come to the Jordan, and the feet of the priests bearing the ark were dipped in the brink of the water...the waters coming down from above stood and rose up in a heap far off, at Adam, the city that is beside Zarethan, and those flowing down toward the sea of Arabah the Salt Sea, were wholly cut off...And while all Israel were passing over on dry ground, the priests who bore the Ark of the Covenant of the Lord stood on dry ground in the midst of the Jordan, until all the nation finished passing over the Jordan. (Josh. 3:14-17)

And what was the reason for this display of the mighty forces of God? To let the people know that Joshua now commanded Israel, and that the Lord was with him also:

And the Lord said to Joshua, "This day I will begin to exalt you in the sight of all Israel, that they may know that as I was with Moses, so I will be with you." (Josh. 3:7)

This was no less than a psychological conditioning exercise through a physical display similiar to that of the Red Sea incident and a confidence-building scheme to ferment a solid faith in the chosen proxy leader.

It is interesting to note too that the Ark remained in the river bed while Israel passed over. The Ark thus becomes the focus of the event. Whatever forces are present in or around the Ark, the people had to be protected from it, and they are warned to maintain a two thousand cubit distance (1,000 yards, or about ten football fields) from it at all times. The fact of the power inherent in the Ark is underlined, as in other references which indicate that one of the duties of the Levite priests was to encamp around the Ark to protect *their own people* from its mysterious powers:

The people of Israel shall pitch their tents by companies, every man by his own standard, but the Levites shall encamp around the tabernacle of the testimony, that there may be no wrath upon the congregation of the people... (Num. 1:52-53)

One of the first great victories of the conquest was the capture and destruction of Jericho. The story of how Joshua led his troops around the walls of the great fortress, the seven priests blowing their trumpets, followed by the ark and the tribes of Israel, is familiar to many. After circling the structure for seven days with trumpets blowing continually as they marched around, the walls began to topple, and the occupants were exposed to the Israelites' advance.

One explanation advanced for the toppling of the walls of Jericho by archaeologists is that Jericho rested in an earthquake zone and an unusually strong tremor could have leveled the fort. (133) It is understandable that secular historians and biblical harmonizers would opt for such an explanation; what other terrestrial explanation is there? But there are problems. Did a devastating tremor suddenly occur at precisely the moment Joshua was preparing to give siege? If so, that would be quite a coincidence! Are we to assume that the fall of Jericho really did not occur at all? Clearly to dismiss the straightforward report of the destruction of Jericho would seem to undermine the credibility of other biblical narratives as well as archaeological evidence supporting such an event. But an extraterrestrial perspective adds the needed dimension to allow for a literal translation on a secular level. Just prior to their march around the fortress, we learn that the Starmen are present:

When Joshua was by Jericho, he lifted up his eyes and looked, and behold, a man stood before him with his drawn sword in his hand; and Joshua went to him and said to him, "Are you for us, or for our adversaries?" And he said, "No; but as commander for the army of the Lord I have now come." (Josh. 5:13-14)

In a rare appearance, this Starman was probably doing what any commander would do: get a first-hand assessment of the situation from the ground where the event was to take place.

Again the Ark plays a prominent role in the objectives of the Starmen. On one level, we are led to believe that it was some power associated with the Ark, either directly or in conjunction with some other factor, such as the pillar vehicle, which caused the walls of the ancient fort to fall. On another level, the Ark was the ceremonial focus of the people. For six days the army marched around the fort in a solemn procession, once each day. On the seventh day, they marched around the fort seven times in a row, the priests marching in front of the Ark, blowing their trumpets continually. As they finished the ritual march, Joshua commanded the army to shout at the tops of their voices:

So the people shouted, and the trumpets were blown. As soon as the people heard the sound of the trumpet, the people raised a great shout, and the wall fell down flat; so that the people went up into the city, every man straight before him, and they took the city. (Josh. 6:20)

It obviously wasn't the shouting which brought the walls down. Nor was it the trumpets. Perhaps the finale to the ritual was calculated to both mask the Starmen's method of bringing down the walls and to pump the warriors into a battle frenzy for the subsequent raid on the fort. We know the Starmen were present, and we know from Joshua 5:13-14 and other passages that the Israelites were not acting alone, but had an *army* of Starmen behind them. Accordingly, the same force which stopped the flow of the Jordan River and parted the Red Sea would have been sufficient to topple the walls of such a fort, and this, therefore, is the logical explanation for the destruction of the walls of Jericho.

There are many battles recorded in the Old Testament. Israel seems to have fought most of its own battles with the Starmen providing more discrete assistance, such as reconnaissance and strategy information as required. But what happens when the Lord decides the situation is desperate or important enough to intervene directly? A vivid account is given of such a situation in a battle against the Amorites at Gibeon.

And the Lord threw them into a panic before Israel, who slew them with great slaughter...And as they fled before Israel, while they were going down the ascent of Bethheron, the Lord threw down great stones from heaven upon them as far as Azekah, and they died; there were more who died because of the hailstones than the men of Israel killed with the sword. (Josh. 10:10-11)

The battle raged into the night, but it did not grow dark:

Is it not written in the Book of Jashar? The sun stayed in the midst of heaven, and did not hasten to go down for about a whole day. There has been no day like it before or since... (Josh. 10:13-14)

There arises no particular difficulty over the technical feasibility of lighting a battlefield. Today we employ mortar-fired flares, such as the 81 mm. M301A3 used by the Marine Corps, which burns with an intensity of 500,000 candle power to light an entire battle area. A large illumination searchlight from the pillar vehicle hovering overhead could have served the same purpose. "Hailstones" and great "stones from heaven" are obvious references to bombs, which were so devastating they killed more of the enemy than Israel killed in contact fighting.

It should by this time be apparent that Israel's hailing of their Lord as a mighty "man of war" was a literal compliment. There is reference to a book entitled *The Book of the Wars of the Lord,* (Num. 21:14) which, like several other biblical books, has long since been lost or abandoned. But if it had survived, how might it have changed our perception of God?

Israel was now well into its reorientation with a new set of laws, new rituals, a military organization, and a sense of purpose. We would expect that the mission of the Starmen was also well on its way to being realized. But there emerges evidence of increasing problems.

For all of its acquired sophistication, Israel constantly had problems breaking with old superstition and silly pagan idols. Despite the incredible, overt, physical displays by the Lord, the constant drilling of the ritual and the harsh punishments meted out for backsliding, many members persisted in their folly. This greatly angered the Starmen and caused them no end of frustration.

Although many cities had by this time been conquered, there remained in Canaan strongholds such as Gezer and Jerusalem still to be overcome before the land was entirely theirs. But Joshua, after years of fighting, was becoming advanced in years himself, and there appeared no way to complete the conquest before his death.

To complicate matters, there seemed to be no suitable successor for Joshua, which meant only one thing: the conquest would have to be halted temporarily. For a time Israel would have to live under the threat of foreign influence and invasion, which concerned the Starmen. The proxy Joshua is instructed to gather his nation together and make final appeals to them: they must not turn away from their God and worship other deities nor forget the laws of Moses:

> *But Joshua said to the people, "You cannot serve the Lord; for he is a holy God; he is a jealous God; he will not forgive your transgressions or your sins. If you forsake the Lord and serve foreign gods, then he will turn and consume you, after having done you good."* (Josh. 24:19-20)

Paradoxically in Joshua's words is a warning to the people not to turn aside from their ordinances, and yet they could not "serve the Lord." What he meant was that the people could not administer directly to the Lord, for that was the duty of the Levites and a potentially dangerous job. But they must keep their laws and not adulterate their new order with foreign ways or idols.

Finally Joshua dies at 110 years of age, and Israel enters into a relatively stagnant period under the leadership of various lesser leaders called "judges." According to a passage in the *Book of Judges,* "Every man did that which was right in his own eyes" (Judges 17:6) during that time. And while it appears there may have been limited Starman activity, no significant action was recorded for this time. The conquest ground to a halt.

* * *

And the Lord visited Hannah, and she conceived and bore three sons and two daughters. And the boy Samuel grew in the presence of the Lord. (I Samuel 2:21)

Samuel became the last in the line of thirteen judges who presided over Israel in about the eleventh century B.C. As a boy he worked under the priest Eli, who was also the twelfth judge over Israel, at Shiloh, where the ark was kept for a time.

Now the boy Samuel was ministering to the Lord under Eli. And the word of the Lord was rare in those days; there was no frequent vision. (I Sam. 3:1)

One day while Samuel was lying down in the temple where the Ark was kept, he heard a voice calling his name, *"Samuel! Samuel!"* (I Sam. 3:4) It had been many years since the last communication through the Ark, and so Samuel's first reaction was that Eli was calling, and he immediately went to him. But Eli had not called. The voice called out again, and again Samuel went to his master. Then it dawned on Eli that the voice had come from the Ark, so he positioned the boy before it and instructed him on how to answer. When the voice returned, Samuel was to reply, *"Speak, for thy servant hears."* (I Sam. 3:10)

Then the voice came again, calling, and Samuel answered. Suddenly after some 200 years of silence, there was again contact with the Lord, and He began to speak to Israel through Samuel concerning the period presided over by the judges. The verdict was grim: Israel would be punished for not adhering strictly to the teachings of Joshua and Moses. They had faltered badly during the absence of the Starmen, greatly corrupting themselves and engaging in practices abominable to the Lord.

And what was the punishment? Israel was surrounded by foreign powers eager to acquire some or all of their prime real estate. It was only a matter of time before an army greater than themselves would challenge them. The Starmen would therefore simply withdraw military assistance for as long as necessary to convince the Israelites that they were doomed without "divine" aid. It did not take long for this scenario to develop.

When the Philistines dealt Israel a humiliating defeat at Aphek, the Hebrew leaders foolishly decided to risk bringing the Ark from Shiloh to invoke its great powers as in the days of old. But it was a vain undertaking, for the Ark was useless without the Starmen to respond to their calls. The Ark was then captured in the second battle.

But it was also a mistake for the Philistines to take this device, of which they had no understanding. Without proper handling by trained attendants, they would find themselves suffering some of the devastating effects which the Lord so often warned of. And so it was:

The hand of the Lord was heavy upon the people of Ashdod and he terrified and afflicted them with tumors, both Ashdod and its territory. (I Sam. 5:6)

Attempts to move the Ark to Gath produced similiar tumors on "young and old." So they tried a third place, with the same results. Finally they decided that they had had enough. There was terror at Ekron; people were dying:

For there was a deathly panic throughout the whole city. The hand of God was very heavy there; the men who did not die were stricken with tumors, and the cry of the city went up to heaven. (I Sam 5:11-12)

Of course the Ark was humbly returned to Israel straightaway. Then, as if to underline the indiscriminate powers of the Ark, it is revealed that when the Israelites received it back and crowded around and tried to inspect it, seventy of their own men were killed!

It seems the Ark was as much a problem as it was indispensable to Israel. Several years later the great warrior David was preparing to enter Jerusalem triumphantly with the Ark when an unfortunate accident occurred:

And when they came to the threshing floor of Nacon, Uzzah put out his hand to the Ark of God and took hold of it, for the oxen stumbled. And the anger of the Lord was kindled against Uzzah; and God smote him there because he put forth his hand to the Ark; and he died there beside the Ark of God. And David was angry because the Lord had broken forth upon Uzzah; and that place is called Perezuzzah to this day. And David was afraid of the Lord that day... (II Sam. 6:6-9)

So distressed was David by this incident that instead of taking the Ark triumphantly into the city as he had planned, he decided that it should remain far away from the masses, and so he took it to a house on the outskirts, where it stayed for three months.

About the time of Samuel, the priest and the prophet, Israel was under military stranglehold from the Philistines on several fronts and was in dire need of a strong leader and warrior to break the pressure and turn the tide. To this challenge rose Israel's first king, Saul, who reigned from c.1,020 to 1,000 B.C. (134) Though Saul fought many valiant battles, no real progress was made until the reign of Israel's second king, David.

David, well known to many as the boy who slew the huge Goliath, grew into a man of many talents. Besides being a first-class warrior, he was an accomplished musician, statesman and poet. He ruled from 1,000 B.C. until 961 B.C., and it was under him that Israel finally gained its homeland in Canaan, thus fulfilling the Lord's long awaited promise.

While his attributes were diverse and notable, our interest lies in his contacts with the Starmen, and our most important clues regarding this relationship are to be found in one of his songs. Although presented previously, it is worth repeating this song sung by David after the defeat of the Philistines at Gob so that we may bring together what has been learned thus far about the methods behind the Starman mission:

In my distress I called upon the Lord; and to my God I called. From his temple he heard my voice, and my voice came to his ears.

We can now understand that David, in the heat of a battle, called upon the Starmen through the Ark, and the appeal was received at their base in the mountains.

Then the earth reeled and rocked; and the foundations of the heavens trembled and quaked, because [the Lord] was angry.

Smoke went up from his nostrils, and a devouring fire from his mouth; glowing coals flamed forth from him.

He bowed the heavens and came down; thick darkness was under his feet. (II Sam. 22:8-10)

The Starmen launched a vehicle immediately, and it arched from the mountain base to the battle area. As it lifted off, it felt as if the whole earth were quaking. Behind the vehicle shot a plume of fire and smoke from the engines. Note how the effects described above are similar to the launch of the space shuttles, and consider how awesome such a display would be to the primitive people of the Old Testament.

He rode upon a cherub and flew; he was seen upon the wings of the wind.

He made darkness around him his canopy, thick clouds, a gathering of water.

Out of the brightness before him, coals of fire flamed forth. (II Sam. 22:11-13)

Here we have an extremely graphic and vivid description of the vehicle slowing to a hover over the battle area with careful attention given to distinguishing between the Lord, his vehicle and its attendant effects.

The Lord thundered from heaven, and the Most High uttered his voice.

And he sent out arrows, and scattered them; lightnings and routed them.

Then the channels of the sea were seen, the foundations of the world were laid bare, at the rebuke of the Lord, at the blast of the breath of his nostrils. (II Samuel 22:14-16)

The enemy was quickly routed with rockets and lasers, or some similiar devices, and where these landed, the "channels of the sea were seen" and "the foundations of the earth laid bare" because of the terrific blast effects. Israel thus claimed a magnificent victory that day.

The tactic employed is familiar to most military people today as *forward air control* and *close air support* used in modern assaults on fortified targets. When troops on the ground encounter a pocket of stiff enemy resistance, swift jets and aircraft are called in to bomb, rocket and strafe enemy positions according to directions given over a radio by a forward air controller on the ground. Once the enemy is sufficiently weakened by the strikes, the troops again move forward and finish the job. Stated poetically in Isaiah, this practice was often used by Israel in its conquest of Canaan:

For thus said the Lord to me, "As a lion...growls over its prey, and when a band of shepherds is called forth against him [and] is not terrified by their shouting or daunted by their noise, so the Lord of hosts will come down to fight upon Mount Zion and upon its hill. Like birds hovering, so the Lord of hosts will protect Jerusalem; he will protect and deliver it, he will spare and rescue it." (Isaiah 31:4-5)

It should now be apparent that the Ark was important to Israel for this reason: the key advantage of the Israelites over other large armies was their ability to call "air strikes" through the Ark. In this way they could maintain a high degree of mobility during the exodus and the conquest while minimizing exposure to the Starmen.

What about weapons used to rout the enemies? The lasers descibed as "lightnings" and the projectiles described as "arrows" have many parallels in other parts of the Bible. Recall that in the Garden of Eden we were told of the "flaming sword" wielded by the cherubim who guarded the entrance to the compound. (Gen. 3:24) At Sodom and Gomorrah there was the "brimstone and fire" which fell from heaven, destroying the cities and Lot's wife. (Gen. 19:24) As Jacob journeyed with his family and all his possessions to his new home in Bethel, a "terror from God" fell upon the cities round about to protect him from his pursuers. When the people initially complained of their lot near the beginning of the exodus, the Lord produced an intimidating display of "fire" that consumed the outlying areas of the camp. (Num.11:1) During the battle with the Amorites, Joshua was aided by the Lord, who "threw down great stones from heaven" upon them near the ascent of Bethhoron. More died because of these "stones" than Israel killed in battle. In Psalms 7:13 His lasers are called "fiery shafts." Similarly "fiery arrows" and "lightnings" can be found in this context in 2 Samuel 22:15, 2 Kings 1:10, Job 36:30 and Psalms 18:14.

What about the effects of these weapons? We know that they were powerful enough to destroy armies and cities. Apparently some survived to tell of the weapons' effect on them:

The stouthearted were stripped of their spoil;
they sank into sleep
all the men of war
At thy rebuke, O God of Jacob,
both rider and horse lay stunned.
(Ps. 76:5-6)

From Lamentations, the account of a victim:

From on high he sent fire;
into my bones he made it descend;
he spread a net for my feet;

he turned me back;
he has left me stunned,
faint all the day long.
(Lam. 1:13)

So, without much interpretation, we find weapons employed by the Starmen capable of producing a variety of effects from stunning to killing and mass destruction. They appear quite similar to weapons we already use or are developing, and they were used precisely the way we would expect weapons to have been used by intelligent human-like aliens so long ago.

* * *

As much as King David did for Israel and for the objectives of the Starmen, there is evidence that he may have overstepped his authority with the Lord on at least one occasion. In I Chronicles there is a brief but interesting account of how David instructed his commander-in-chief to go about Israel with his troops and take a census of the people. This appears to have so infuriated the Starmen that the Lord sent an angel to Jerusalem to destroy it!

So the Lord sent a pestilence upon Israel; and there fell seventy thousand men of Israel. And God sent the angel to Jerusalem to destroy it; but when he was about to destroy it, the Lord saw, and he repented of the evil; and he said to the destroying angel, "It is enough; now stay your hand." And the angel of the Lord was standing by the threshing floor of Ornan the Jebusite. And David lifted his eyes and saw the angel of the Lord standing between earth and heaven, and in his hand a drawn sword stretched out over Jerusalem. (I Chron. 21:14-16)

Almost more incredible than the severity of the punishment is the characterization of the Lord as *repenting* of the "evil" he sought to deal David's people! While the act of taking a census may in fact have been only the "final straw that broke the camel's back" in a series of acts by David which were found mutinous by the Starmen, it is nevertheless important to note the fit of anger and subsequent reversal of an important decision by the Lord. He was furious that despite all of His demonstrations of power and authority, with traditions dating back to Abraham and further, that the people of Israel continued to stray, to backslide and even to disobey.

To reinforce the evidence of these trends, we return briefly to the exodus, to the account which describes the Lord's wrath against Israel for making idols. When Moses went up to Mount Sinai to receive the Ten Commandments, the people waited and waited and began to think Moses had been killed or lost because he was gone so long. The Lord instructed Moses to return to his people for they had in his absence constructed the likeness of a calf, a common animal of worship for Semitic peoples at that time (135), and had called upon it to deliver them out of their plight in the wilderness. The Lord was so incensed at the rapidity with which the people had turned from his instruction, He vowed to wipe them out:

And the Lord said to Moses, "I have seen this people, and behold, it is a stiff-necked people; now therefore let me alone that my wrath may burn hot against them and I may consume them; but of you I will make a great nation."

But Moses besought the Lord his God, and said, "O Lord, why does thy wrath burn hot against thy people whom thou hast brought forth out of the land of Egypt with great power and a mighty hand? Why should the Egyptians say, 'With evil intent He did bring them forth to slay them in the mountains and to consume them from the face of the earth'? Turn from thy fierce wrath and repent of this evil against thy people."

And the Lord repented of the evil which he had thought to do to his people. (Ex. 32:9-12, 14)

Perhaps, in addition these two examples will illustrate for the reader another important dimension of the relationship between the Starmen and their chosen people: the Hebrews were indeed a "stiff-necked" people! It is almost incredible that they found it so easy to let themselves fall prey to pagan idols and abominable foreign practices in the face of such clearly stated consequences. The pressures upon them both from the Starman indoctrination and from foreign influences must have been nearly equally great; as the influence of one waned, the other waxed strong upon them. On the other hand, the Starmen must have experienced tremendous frustrations over this alien "kin" which they were trying so persistently to tame. From the fall of the first hybrids at Eden to the rejection of Christ, the Bible is consistent in recording this conflict between the objectives of the Lord and the weakness, almost rebelliousness, of his people in realizing them.

And yet there did exist a basic respect on both sides which began to mature into something extraordinary. It may have begun when Israel finally began to settle down in Canaan after years of preparation in the wilderness. Finally, Israel was to enjoy a generation of peace and prosperity under a new king selected by the Lord and born to David:

Behold, a son shall be born unto you; he shall be a man of peace, I will give him peace from all his enemies round about; for his name shall be Solomon, and I will give peace and quiet to Israel in his days. He shall be my son and I shall be his father, and I will establish his royal throne in Israel forever. (I Chron. 22:9-10)

Solomon's nick-name was Jedidiah, the son of Bathsheba, who, under the "pure vine" metaphor, was probably raised up by the use of artificial insemination to be the actual son of the Lord. He was annointed King in 961 B.C. and ruled over Israel for about forty years. At no other time in Israel's history had there been so much prosperity, nor was the territory held by Israel ever so extensive as under the wise and worldly Solomon. Much of the credit for Israel's peace can be attributed to the weakened states of Egypt and Assyria during that period, and Solomon took advantage of the condition to make his kingdom glitter with pomp and opulence.

As well as an accomplished poet and naturalist, Solomon was a businessman extraordinaire. He forged new trade alliances, added new trade routes by land and sea, and built a chain of war chariot cities to defend the newly acquired wealth. (136)

He also built a lavish temple, a permanent resting place for the Ark of the Covenant, according to a plan which was given to his father, King David, by the Lord. The Ark now no longer needed to be carried about by the Levite priests. It had fufilled its primary mission during the exodus as a communications device and would now presumably assume a primary role as a ritual instrument.

* * *

Shortly after the reign of King Solomon ended, the great kingdom which was attained through so much sweat and blood began to disintegrate, not from without, but from within. Perhaps it was the great wealth and cosmopolitanism cultivated by its business and culture-conscious king which brought about the loosening of the sacred values of their Lord. Perhaps the struggle for the throne after Solomon's demise was the ultimate cause. In any case, the kingdom soon split into two rival states, with Judah in the south and Israel in the north, and was ruled by two kings until each was conquered by foreign armies only a few centuries later.

It is also about this time that we lose track of the Ark. The complex historical aspects of this period are beyond the scope of this book, but it may be noted that no evidence for the existence of the Ark of the Covenant is to be found after the Babylonians begin the conquest of Jerusalem in 597 B.C. and destroy the temple of Solomon and its contents.

Had the Ark been captured by the Babylonians and destroyed along with the other temple artifacts, or had Jeremiah the priest been able to hide it? It is evident that the Starmen knew about the invasion. They probably reclaimed it rather than have it fall into unauthorized hands. With all the reported troubles experienced by the Philistines when they captured the Ark, it is to be expected that a similar fate would have befallen the Babylonians had they actually taken it. As late as the reign of Judah's King Hezekiah, (c. 715-687 B.C.) the Ark was still in use (Is. 37:16), but during the reign of King Josiah (c. 640-609 B.C.), a mere four years before the first siege of Judah, the Ark is spoken of in the past tense:

> *"And when you have multiplied and increased in the land, in those days," says the Lord, "they shall no more say, 'The ark of the covenant of the Lord.' It shall not come to mind, or be remembered, or missed; it shall not be made again."* (Jer. 3:16)

We now arrive at another major turning point in the history of Israel. Politically splintered and weakened, Israel had by this time passed the heyday of the King David - King Solomon era. Their once great kingdom, now divided, was slowly being eroded by foreign invasion and by infighting among their own people. Soon the Israelites would again find themselves under foreign domination after so many years of struggling to gain their independence. Despite the intensive reorientation by their god, despite the lofty goals, the bloody battles and the awesome wonders worked by the Lord on their behalf, they were losing their bid to become a nation of priests unto the world, and the reason may have been in part due to the way in which the Starmen sought to win over their kin.

* * *

It is interesting to note that between the time of the division of the Hebrew kingdom and the fall of Jerusalem in 587 B.C., there arose the prophets, or messengers of God, appearing at roughly the three most important points of the period. Less than a century after the kingdom split came the prophet Elijah and after him Elisha. About 20 years before Israel fell to the Assyrians in 722 B.C., there were Amos, Hosea, Isaiah and Micah. Finally the prophets Jeremiah and Ezekiel preceded the fall of the southern kingdom of Judah in 587. Two of the prophets—Ezekiel, and to a lesser extent, Elijah—are reported as having significant contacts with the Starmen during this period.

Shortly after the division of Solomon's kingdom, when there was a high degree of internal strife, the prophet Elijah was summoned by the Lord from the northern kingdom of Israel to try to persuade the people not to turn from the teachings of Moses. But Israel's King Ahab had married a Phoenician princess, and the Phoenician cult worship of Baal Melkarth was allowed during his reign. When Elijah killed some of Baal's prophets, his life was endangered, and he fled into the wilderness, to Mt. Horeb, to seek the Lord and to report the atrocities. In an incident very similar to the contact with Moses some four centuries earlier, we are struck with the image of a large alien ship as it approaches Elijah's cave:

> *And behold, the Lord passed by, and a great and strong wind rent the mountains, and broke in pieces the rocks before the Lord, but the Lord was not in the wind; and after the wind an earthquake; and after the earthquake a fire, but the Lord was not in the fire; and after the fire a still small voice. And when Elijah heard it, he wrapped his face in his mantle and went out and stood at the entrance of the cave. And behold, there came a voice to him, and said, "What are you doing here Elijah?"* (I Kings 19:11-13; compare Exodus 33:18-23)

Just as in the "burning bush" incident, the Lord was understood to be manifest not in the effects produced by the vehicle, but as a separate entity which was inside this thing which Elijah had no adjectives to describe. He is a good observer. He notes the effects of the approaching vehicle and states that the Lord was not any of them. From the vehicle came a "still small voice," as if it were from a loudspeaker on the vehicle and not some intimate inner voice.

But the most interesting aspect of Elijah's life is that he did not die on Earth but, like Enoch, was taken up into heaven:

Now when the Lord was about to take Elijah up to heaven by a whirlwind, Elijah and Elisha were on their way from Gigal ...And as they still went on and talked, behold, a chariot of fire, and horses of fire separated the two of them. And Elijah went up by a whirlwind into heaven. And Elisha saw it and cried, "My father, my father!" And he saw him no more. (II Kings 2:1, 11-12)

There simply was no way for this primitive man to comprehend that an exotic machine, flown by a man, could swoop down from the heavens, scoop up another man and carry him up into "heaven" never to return! In a display that was at once both frightening and majestic, the Starmen had claimed one of their faithful watchers and disappeared into the sky.

Because of the corruption and turmoil in the northern kingdom, it fell to the Assyrians not long afterward. Still, there was the kingdom of Judah to the south, and as long as there were faithful kin left, an effort had to be made to salvage them.

Fall of Jerusalem

It becomes evident in the study of the Bible that it is a story of an epic journey of a people from their unique beginnings in Eden through centuries of trials and tribulations at the hand of their Lord to the gaining for themselves a place on earth they could call their own. But while the goal of the people may have focused on gaining a homeland, the goal of their Lord went further, toward making them teachers and models for the rest of mankind. Perhaps because of this misunderstanding of ultimate goals, however, toward the latter parts of the Old Testament it is also the chronicle of their fall from grace, their ruin and destruction.

It is near the end of the seventh century B.C. that the Starmen begin to realize that they are irretrievably losing their created people whom they had so jealously guarded and nurtured across the vast expanse of time since Eden. It was during this tumultuous period preceding the conquest by Babylon that they begin to express a great sense of loss and dismay at the outcome of their project, and this emotion is reflected in a number of passages attributed to the Lord, especially in the books of Jeremiah, Ezra and Isaiah:

"You have rejected me," says the Lord, "You keep going backwards..." (Jer. 15:6)

"Yet I planted you a choice vine, wholly of pure seed. How have you turned degenerate and become a wild vine?" (Jer. 2:21)

"The people of Israel and the priests and the Levites have not separated themselves from the peoples of the lands with their abomination, from the Canaanites, the Hittites, the Perizzites, the Jebusites, the Ammonites, the Moabites, the Egyptians and the Amorites. For they have taken some of their daughters to be wives for themselves and their sons; so that the holy race has mixed itself with the peoples of other lands." (Ezra 9:1-2)

Hear O heavens and give ear O earth, for the Lord has spoken; "Sons have I reared and brought up, but they have rebelled against me. The ox knows its owner and the ass its master's crib; but Israel does not know, my people does not understand." (Is. 1:2-3)

It was imperative for the Israelites to maintain their cultural and racial integrity. They were not to intermarry among other peoples or mix cultures. To the Starmen, Israel was like a line of purebred animals, carefully bred and reared the way we prize select strains of horses and dogs, etc. Such a mentality is comprehensible, if unsavory, considering the extreme social, cultural and technical disparity between the Starmen and their hosts. The expectation of the Starmen seems to have been that since they created and raised the Hebrews, they could demand obeisance in return for their efforts. When the Hebrews corrupted themselves, and continued to do so, the Lord was greatly distressed and perplexed.

So much energy, time and material had been spent on the project. Not only did they at first not realize what they had done wrong, they were at a loss about what to do at this point. The Starmen were at once dismayed and angry, again confirming their human natures and further supported by the Lord's use of the interrogative: *"How have you turned degenerate...?"* (Jer. 2:21) Moreover, the Lord's complaint that Israel keeps "going backwards" confirms a plan for the Hebrews, a plan which was not working. The Starman mission was failing, and they began to ask themselves why.

The Starmen had kept their promise: Israel, if only for a brief time, had enjoyed the prosperity and security of a homeland in Canaan. But soon after Solomon's death, the great kingdom fell apart, and the Starmen were stung by their setback. Rather mysteriously, sometime during the reign of Solomon, they seem to have again withdrawn to a very low-key position and perhaps had even left for a time.

From the masses there arose a great sense of anxiety over the lack of stability in their goverment, and dark clouds of insecurity seem to overshadow their literature during this period just prior to the invasion of Jerusalem. The elders of Israel began to lament the absence of the Lord and to tell the younger generation of His great works in days gone by. Hearing of this, the people begin to mourn the loss of their God:

We have heard with our ears, O God, our fathers have told us, what thou didst perform in their days, in the days of old; thou with thy own hand didst drive out the nations, but them thou didst plant. (Psalms 44:1-2)

O God, why doest thou cast us off forever? Why does thy anger smoke against the sheep of thy pasture?...direct thy steps to the perpetual ruins; the enemy has destroyed everything in the sanctuary...they set thy sanctuary on fire; to the ground they desecrated the dwelling place of thy name. (Psalms 74:1, 7)

Will the Lord spurn forever, and never again be favorable ...are his promises at an end for all time? (Psalms 77:7-8)

It had been a long time in the making, the fall of Jerusalem. Caught between the two "superpowers," Egypt and Babylon, it was only a matter of time before the divided and weakened Hebrew nation became a tempting target. Finally it happened:

At that time the servants of Nebuchadnezzar king of Babylon came up to Jerusalem, and the city was besieged...and Jehoiachin the king of Judah gave himself up to the King of Babylon, himself, and his mother and his servants and his princes and his palace officials. The King of Babylon took him prisoner in the eighth year of his reign and carried off all the treasures of the house of the Lord, and the treasures of the king's house, and cut in pieces all the vessels of gold in the temple of the Lord, which Solomon, King of Israel had made, as the Lord had foretold. He carried away all of Jerusalem, and all the princes and all the mighty men of valor, ten thousand captives...none remained except the poorest people of the land. (II Kings 24:10-14)

The Starmen would no longer fight for the Hebrews. They found themselves immersed in a host of problems for which they had no immediate answers, and they seem to have been allowing their great mission to slide into oblivion after centuries of effort. What had gone wrong?

There are, of course, *innumerable* reasons why such a mission could have failed. One of the most obvious seems to have been that the Starmen's approach was just too impersonal, too awesome and too frightening to the relatively primitive Hebrews. Instead of developing close interpersonal relationships with people, the Starmen dealt with them through proxies, and any devotion shown to the Lord under those circumstances was primarily due to a respect based on fear. While this fear and respect of an abstract God facilitated a degree of psychological control when the Starmen were present, almost as soon as they departed, the people returned to the more familiar pagan idols of terrestrial cultures, which were just as potent as the Lord when the Starmen weren't present.

A second significant factor may have been that the Starmen came and went throughout Israel's history. Several passages, such as *"Seek the Lord while he may be found, call upon him while he is near..."* (Is. 55:6), make it clear that the Starmen simply are not present during some periods. By leaving the people alone with their relatively new precepts among other cultures with a broad spectrum of magic and religious traditions, it would have been quite difficult, if not impossible, for an entire nation to maintain its integrity for long. To add to the Hebrew's confusion, the Starmen, as gods, besides being just as impotent as other pagan idols when absent, had no form to venerate, to focus on. After a few generations of the Starmen's absence, it would be expected that the people would look to gods which they could at least see and touch. In fact other cultures' gods must have appeared to have been more powerful than the Lord when there was no Lord to protect Israel and other empires were engulfing them from all sides!

A third significant factor was that the reorientation process involved a profusion of new and complex ordinances, values and concepts which the people seem to have found difficult to interpret, understand and assimilate. For example the highly complex ritual associated with the Ark of the Covenant and the tabernacle caused no end of confusion. Throughout the Ark's existence among the Israelites, there was death and destruction because generation after generation continued to misunderstand the forces at work within it. One aspect of the ritual was intended to allow the Israelites to maintain the Ark in their presence with maximum safety. Yet throughout its history, a string of Ark-related mishaps plagued Israel. The masses, it seems, never really were able to share the Starmen's vision for them.

And so it was Israel came once again under bondage. But even during those hard times when the oppression was great, there remained those who retained a seed of hope that their God would someday return to reclaim them.

The Starmen seem to have undergone a dramatic philosophical change at this juncture. They realized that their plan was not working under the existing conditions: the people were alienated; they were resisting the assimilation of the laws of Moses and seemed to have lost sight of their purpose and debt. Was there any way to salvage the operation and any of their people? History reveals that the answer is both "yes" and "no," for basic tenets of the Intermediate Order have been preserved and integrated into later Christianity, but only after the Hebrews forfeited their sole claim to them. What was this new approach? It involved a radical new covenant:

"But this is the covenant which I will make with the house of Israel after those days," says the Lord; "I will put my law within them, and I will write upon their hearts..." (Jer. 31:33)

"...their prince shall be one of themselves, their ruler shall come forth from their midst." (Jer. 30:21)

"In those days I at that time will cause a righteous branch to spring forth from David." (Jer. 33:15)

This was to be quite a different approach from that of the overt, wrathful, disciplinarian god who historically dealt with Israel. It was a new plan by which the Lord hoped to supplant old methods based upon the "fear factor." An ingenious, subtle new technique of psychological manipulation was in the works: control through conscience.

There was to be a time limit for the Hebrews, a period of reflection lasting 490 years, after which a new "branch" would arise to deliver Israel:

...while I was speaking in prayer, the man Gabriel, whom I had seen in the vision at the first, came to me in a swift flight at the time of the evening sacrifice. He came and he said to me, "O Daniel, I have now come out to give you wisdom and understanding ...Seventy weeks of years are decreed concerning

your people and your holy city, to finish the transgression, to put an end to sin, and to atone for iniquity, to bring everlasting righteousness, to seal both vision and prophet, and to anoint a most holy place." (Daniel 9:21-22, 24)

They were thus provided several generations of time to study the scriptures and to prepare themselves once again to become a great nation or to perish under the weight of their sins. Of course the new ruler from their midst was to be Jesus, the son of God. His influence upon the world is obvious, and yet there is a great paradox about Jesus' mission and his impact upon history. Jesus was a Hebrew who was sent by the Lord to claim his kin, but the *whole world* inherited his teachings. We will examine the significance of Jesus more closely in a later section, but first there are some notable events concerning the Starman mission during the period between the fall of Jerusalem and the arrival of the Christ.

And they answered the angel of the Lord who was standing among the myrtle trees, "We have patrolled the earth, and behold, all the earth remains at rest." (Zechariah 1:10-11)

Ezekiel's Vision

When Erich Von Daniken's *Chariots of the Gods?* was first introduced in the United States in the early seventies, it sent shock waves through religious, historic and scientific circles. The idea that alien beings could have come to the earth in the dim reaches of our past to influence our development was no less than revolutionary. What was so shocking was that, if true, it would rend the very fiber of our religious and historical perceptions of humans as the supreme expression of God's creation, and it immediately provoked a volley of essays questioning or defending established beliefs on the subject.

And so it happened that a man named Joseph F. Blumrich, like countless others, had picked up a copy of *Chariots of the Gods?,* and begun to read with amusement what all the fuss was about until he came upon references to the *Book of Ezekiel* and Ezekiel's "visions" of God. Von Daniken's assertion that the visions could have been reports of ancient astronauts and their vehicles caused Blumrich to pause.

As chief of the Systems Layout branch of NASA, Blumrich had worked on spaceship design and construction since his arrival in the U.S. in 1959 and had been instrumental in the development of Skylab and the Saturn V rocket. In his field the object is to distill concepts of a desired space system down into a real working machine - an enormous undertaking considering that many projects involved cutting-edge technology and innovative design. If Von Daniken was so bold as to make such wild assertions about these biblical visions, thought Blumrich, he could easily be proven wrong by applying established principles of rocket and spaceship design. It was an intriguing challenge.

In his spare time, he set about conducting research into the *Book of Ezekiel,* its setting, author, editors and message, etc. But what he discovered, to his amazement, was that the accounts historically regarded as "visions" were more like direct observations by a very competent observer. These accounts were, in fact, so detailed and consistent that Blumrich found it possible to extract relevant data for a competent analysis under Von Daniken's ancient astronaut hypothesis. His conclusion: the *Book of Ezekiel* is indeed the record of a series of visitations by beings resembling men who flew in technically advanced vehicles! He was so impressed with what he found that he wrote his own book entitled *The Spaceships of Ezekiel.* (137)

In doing so Blumrich provided us with perhaps the most convincing evidence yet that what was written in the Bible is indeed a record of the presence of extraterrestrials in ancient times. While he made no attempt to analyze or interpret any part of the Bible other than the *Book of Ezekiel,* his conclusions carry significant weight for the Starman Intervention Theory because Ezekiel's observations are quite consistent with other biblical narratives discussed in this book.

Who then was Ezekiel? During the exile in 597 B.C., most of the influential, educated and wealthy Hebrews were removed from Judah and taken to a place called Tel Abib just south of Babylon. There they lived in a colony by themselves near a canal in the Euphrates irrigation system referred to as the River Chebar. (138) Ezekiel was among those exiles.

Educated as a priest in Jerusalem, Ezekiel is considered to have been a knowledgeable man of upper-middle class status, probably in his thirties when he was deported. His elite status is indicated by the conditions of his existence during the exile, where he lived with his family in his own house and generally exercised a high degree of autonomy and freedom of movement considering his situation.

Ezekiel is credited with authoring his own book although there is evidence of tampering by later redactors who were concerned that his writings, though invaluable, did not always coincide with the teachings of the day. Blumrich notes that the text is somewhat adulterated and fragmented at some points so that parts of the account lack unity or coherence, but not to the point of totally obscuring its message. (139)

The book's contents deal with the Lord's word to his people concerning three main topics: the fall of Jerusalem, oracles pertaining to the Lord's dealing with foreign nations, and a detailed plan for rebuilding and restoring Israel after the exile. There are two categories of messages: the "visions" of real contacts with the Starmen and the prophetic material in which Ezekiel was inspired to write or speak to his people concerning the Lord's word. We shall focus on the visions. There are four such "visions" recorded by Ezekiel with an incredible twenty year span between the first vision and the last.

The book opens five years into the exile with a vivid description of the first "vision" approaching Ezekiel out of the north from where he stood by the river Chebar. It is necessary to quote from *Ezekiel* at length here in order to orient the reader to the material and to illustrate three important points: First, note Ezekiel's powers of observation and capacity for detail. His reports of the "visions" are the lucid recollections of an intelligent, sober observer. Second, note the objectivity of his reports. According to Blumrich (140) they demonstrate a careful rejection of subjective or interpretive reasoning. For the most part, Ezekiel only compares what he sees to known phenomena and does not attempt to pass judgment about what it might have been. Finally, note how material the "vision" reported by Ezekiel seems. It does not appear at all like an hallucination, a psychic experience or a figment of his imagination. It is something he reports as physically present, and it has a great impact upon him, both physically and emotionally.

Book of Ezekiel: Chapter I

1:1 In the thirteenth year, in the fourth month, on the fifth day of the month, as I was among the exiles by the river Chebar, the heavens were opened, and I saw visions of God.

1:2 On the fifth day of the month (it was the fifth year of the exile of King Jehoiachin),

1:3 the word of the Lord came to Ezekiel the priest, the son of Buzi, in the land of the Chaldeans by the river Chebar; and the hand of the Lord was upon him there.

1:4 As I looked, behold, a stormy wind came out of the north, and a great cloud, with brightness round about it, and fire flashing forth continually, and in the midst of the fire, as it were gleaming bronze.

1:5 And from the midst of it came the likeness of four living creatures. And this was their appearance; they had the form of men,

1:6 but each had four faces, and each of them had four wings.

1:7 Their legs were straight, and the soles of their feet were like the sole of a calf's foot; and they sparkled like burnished bronze.

1:8 Under their wings on their four sides they had human hands. And the four had their faces and their wings thus;

1:9 Their wings touched one another; they went every one straight forward, without turning as they went.

1:10 As for the likeness of their faces, each had the face of a man in front; the four had the face of a lion on the right side, the four had a face of an ox on the left side, and the four had the face of an eagle at the back.

1:11 Such were their faces. And their wings were spread out above; each creature had two wings, each of which touched the wing of another, while two covered their bodies.

1:12 And each went straight forward; wherever the spirit would go, they went, without turning as they went.

1:13 In the midst of the living creatures there was something that looked like burning coals of fire, like torches moving to and fro among the living creatures; and the fire was bright, and out of the fire went forth lightning.

1:14 And the living creatures darted to and fro, like a flash of lightning.

1:15 Now as I looked at the living creatures, I saw a wheel upon the earth beside the living creatures, one for each of the four of them.

1:16 As for the appearance of the wheels and their construction; their appearance was like the gleaming of a chrysolite; and the four had the same likeness, their construction being as it were a wheel within a wheel.

*1:17 When they went, they went in any of their four directions without turning as they went. (*Another reading is: *When they went, they went on their four sides without turning as they went.)*

1:18 The four wheels had rims and they had spokes; and their rims were full of eyes round about.

1:19 And when the living creatures went, the wheels went beside them; and when the living creatures rose from the earth, the wheels rose.

1:20 Wherever the spirit would go, they went, and the wheels rose along with them; for the spirit of the living creatures was in the wheels.

1:21 When those went, these went; and when those stood, these stood; and when those rose from the earth, the wheels rose along with them; for the spirit of the living creatures was in the wheels.

1:22 Over the heads of the living creatures there was the likeness of a firmament, shining like a crystal, spread out above their heads.

1:23 And under the firmament their wings were stretched out straight, one toward another; and each creature had two wings covering its body.

1:24 And when they went, I heard the sound of their wings like the sound of many waters, like the thunder of the Almighty, a sound of tumult like the sound of a host; when they stood still, they let down their wings.

1:25 And there came a voice from above the firmament over their heads; when they stood still, they let down their wings.

1:26 And above the firmament over their heads there was a likeness of a throne, in appearance like sapphire; and seated above the likeness of a throne was a likeness as it were of a human form.

1:27 And upward from what had the appearance of his loins I saw as it were gleaming bronze, like the appearance of fire enclosed round about; and downward from what had the appearance of his loins I saw as it were the appearance of fire, and there was brightness round about him.

1:28 Like the appearance of the bow that is in the cloud on the day of rain, so was the appearance of the brightness round about. Such was the appearance of the likeness of the glory of the Lord. And when I saw it, I fell on my face, and I heard the voice of the one speaking.

Thus reads chapter one in its entirety. Even to the casual reader, it should be apparent that whatever Ezekiel saw, it was more than a dream. His frequent use of the phrases *"like the appearance of," "such was the likeness of"* and *"they had the form of"* indicate quite clearly that he was attempting to describe something for which he had no experience. It was an utterly alien sight, and all that he could do was to make comparisons with familiar shapes, materials and acoustical phenomena. For this reason some of his descriptions will be immediately familiar to the reader while others may seem a bit cryptic or symbolic.

The things with which Ezekiel had experience—such as fire, lightning, the wheels and the sounds—are easily recognizable to us through his descriptions. When he encounters other aspects of the vehicle, however, the descriptions appear more abstract and will require some interpretation based upon Blumrich's expert opinion. Those interested in a detailed analysis of the *Book of Ezekiel* are encouraged to read Blumrich's excellent book, *The Spaceships of Ezekiel.* Our primary interest is in pointing out how his analysis supports the theory of extraterrestrial intervention in the events of the Bible.

The entire account is testament to Ezekiel's wonderful capacity for observation, attention to detail and understatement, and is easily broken down into component parts for analysis. It will be instructive to evaluate this first "vision" component by component in the light of Blumrich's analysis in order to demonstrate the feasibilty of the hypothesis that what Ezekiel saw was the approach of an extraterrestrial space ship.

The narrative immediately opens with Ezekiel's attention being drawn to the sky, where the silence is shattered by the thundering reentry of an alien vehicle from a polar orbit: (141)

1:4 As I looked, behold, a stormy wind came out of the north, and a great cloud, with brightness round about it, and fire flashing forth continually, and in the midst of the fire, as it were, gleaming bronze.

Anyone who had the privilege of witnessing the spectacular launch of the space shuttle *Columbia* one misty April morning in 1981 would most probably recognize in Ezekiel's report a similar sequence of events only in reverse order:

Trailing a Promethean plume of fire and smoke, the entire 18-story-high, 4.5 million-lb. package thundered off the pad, shaking the earth for miles around. (142)

The major differences in the two accounts are the times in which they occurred and the fact that while *Columbia* was emerging *from* its cloud of smoke, the vehicle seen by Ezekiel was entering the Earth's atmosphere *into* its cloud of smoke and vapor. Instead of seeing the vehicle rising above its plume, Ezekiel could only catch glimpses of the main body of the descending craft as its metallic skin reflected sunlight from amid its exhaust clouds.

The alien vehicle slowed down and came into view, looking nothing like our *Columbia*, nor like anything we have ever produced. It was an advanced design, according to Blumrich, incorporating three main components visible to Ezekiel: the main body with a powerful rocket for entry and exit through the terrestrial atmosphere, four helicopter units (used for short flights at low altitude about the planet), and the capsule at the top of the vehicle and which contained the cockpit.

From Blumrich's analysis, the shape of the main body is deduced as being very similar to a child's spinning top proportioned rather like a mushroom. It had a convex dome at the top, like the cap of a mushroom, and then tapered down quickly to a narrow base resembling the stem of a mushroom, forming a concave lower body. The reason for such a configuration is quite logical and straightforward according to Blumrich: the entire lower surface provided the aerodynamic braking required to slow the vehicle sufficiently during reentry to keep it from burning up due to friction with the upper atmosphere:

The flight from space through the air and to the earth begins with a velocity of some 21,300 miles per hour. For the landing on the surface of the earth this enormous speed must be reduced to zero. By far the greatest part of this braking can be achieved by aerodynamic drag. The quasiconical lower portion of the spacecraft is superbly suited to this objective. (143)

Situated almost directly beneath the outer rim of the mushroom-like upper body of the vehicle, Ezekiel saw what Blumrich calls the "helicopter units." Once the reentry speed of the vehicle is sufficiently reduced, a rocket can be fired at the base of the main body to slow its descent enough to allow for the deployment of these helicopter units. The units move into place at four points on the rim of the vehicle, much like the landing gear on modern aircraft, and provide the final thrust for braking action and for maneuvering to a landing position.

It is immediately after sighting the approaching vehicle that Ezekiel witnesses the deployment of the four helicopter units and describes them as having the approximate form of "men," although he is careful to note important differences:

1:5 And from the midst of it came the likeness of four living creatures. And this was their appearance; they had the form of men,

1:6 but each had four faces, and each of them had four wings.

1:10 As for the likeness of their faces, each had the face of a man in front; the four had the face of a lion on the right side, the four had a face of an ox on the left side, and the four had a face of an eagle at the back.

While the traditional interpretation of Ezekiel's vision classified these mysterious creatures as angels, Blumrich sees an obvious and practical mechanical configuration which served a multi-purpose role in conjunction with the main body of the vehicle.

They seem to have been cylindrical units, extending and retracting from the main body of the vehicle on a single heavy strut. This would be to protect critical parts from the friction of reentry. Blumrich concludes in his analysis that each unit, when deployed, assumed an upright position perpendicular to the ground. In this way the suspended units must have given Ezekiel the impression of torsos.

Blumrich speculates that Ezekiel's description of four faces on each unit were his impressions of mechanical configurations on the units - or even images, like the logos, symbols and characters we paint on contemporary aircraft. Ezekiel remains very cautious throughout, however, taking great pains to compare and not to interpret, so that we must assume he actually saw four faces either molded or painted on the units.

1:11 Such were their faces. And their wings were spread out above; each creature had two wings, each of which touched the wing of another, while two covered their bodies.

1:23 And under the firmament their wings were stretched out straight, one toward another; and each creature had two wings covering its body.

1:24 And when they went, I heard the sound of their wings like the sound of many waters, like the thunder of the Almighty, a sound of a tumult like the sound of a host; when they stood still, they let down their wings.

At the top of each helicopter unit was a four-bladed rotor. In order to facilitate compact storage, the blades folded up or down over the cylindrical units. When extended and turning, the blades came so close together that Ezekiel thought they were actually touching. Indeed, they may have functioned in a synchronous overlap. On the CH-46 Sea Knight helicopter used by the U.S. Navy and Marine Corps, the fore and aft blades are synchronized because they overlap when turning. When stowed on board an assault ship or aircraft carrier, the blades of the helicopter can be folded along the length of the helicopter's body to save precious deck and hangar space. Anyone who has ever been on or near one of these behemoths is also well aware of the terrific "whishing" noise and "thunderous" roar of the jet turbines, as well as the great "whirlwind" created by the downward thrust of the blades while taking off and landing.

1:8 Under their wings on their four sides they had human hands.

At the sides of each helicopter unit there hung a pair of mechanical arms with graspers, having the appearance of human hands, with the obvious purpose of allowing remote manipulation of objects from the cockpit of the vehicle. We use such mechanical graspers on many of our undersea minisubs and also on our moon probes, where it is difficult or impossible to leave the craft to do physical work. Many of them do, indeed, look like human arms and hands.

1:7 Their legs were straight, and the soles of their feet were like the sole of a calf's foot; and they sparkled like burnished bronze.

Taken together with the descriptions of the other parts of the units, it is easy to understand Ezekiel's comparison to human forms. The mysterious "living creatures" even had "legs!" Beneath each cylindrical helicopter unit there extended a pair of shock absorbing struts with round pads at the ends, giving the impression of legs. They were, in fact, probably the main landing gear for the vehicle and were logically situated at the outermost points on the craft for the greatest stability.

Despite the likeness to human forms, however, it is made clear that all the "creatures" behaved alike. This is an important clue that the four units were not "living creatures" at all. They all moved in perfect unison because they were fixed to the rim of the main body:

1:12 And each went straight forward; whenever the spirit would go, they went, without turning as they went.

Although several of their parts had the outward appearance of "living creatures," their true mechanical nature is thus confirmed here. They were not only identical in every respect, they all acted as one in perfect unison with the main body of the vehicle. If the ship went, they all went. Likewise, if one creature went, they all went, without turning as they went.

Scanning the scene, Ezekiel notices several effects produced by the ship which may or may not have been interrelated:

1:13 In the midst of the living creatures there was something that looked like burning coals of fire, like torches moving to and fro among the living creatures; and the fire was bright and out of the fire went forth lightning.

1:14 And the living creatures darted to and fro, like a flash of lightning.

Blumrich believes that two separate phenomena are described in these passages, the first being the effect caused by the glowing nozzle at the base of the main body. At this point the vehicle was slowly approaching under power of the helicopter units, which had taken over for the reentry rocket engine. After the rocket engine was shut down, its extreme temperature would have caused it to glow for a while, and this is what Ezekiel may have called the "burning coals of fire." Blumrich also speculates that the "torches" coming from the midst of the creatures were actually small rocket thrusters, similar to the type used on contemporary rockets for small altitude and course corrections while in flight.

Another possible explanation not mentioned by Blumrich is that Ezekiel was witnessing flashing lights on the alien vehicle, either as a single phenomenon or in conjunction with those mentioned by Blumrich. A special navigation light used on modern aircraft, called an *anti-collision light,* produces an effect like flashes of lightning. As its name implies, the anti-collision light is a visual aid to pilots for spotting other aircraft in flight to reduce the risk of mid-air collisions. Each light assembly is composed of one or more rotating beacons and on our aircraft are brilliant red in color. Most aircraft have these lights on top and bottom surfaces, and when the plane is on the ground, these lights sweep all around in a flashing array. The effects of these beams can also be seen from the window of a passenger plane as it passes among the clouds at night. If Ezekiel's vehicle was coming in for a landing with a similar array of lights flashing among its shiny helicopter units and its vapor clouds, the effect might be very similar to Ezekiel's description, with the shadows of the helicopter units appearing to dart "to and fro" against the exhaust clouds, attended by the flashing of bright "lightnings."

1:15 Now as I looked at the living creatures, I saw a wheel upon the earth beside the living creatures, one for each of the four of them.

1:16 As for the appearance of their wheels and their construction; their appearance was like the gleaming of chrysolite; and the four had the same likeness, their construction being as it were a wheel within a wheel.

*1:17 When they went, they went in any of their four directions without turning as they went. (*Another reading is: *When they went, they went on their four sides without turning as they went.)*

1:18 The four wheels had rims and they had spokes; and their rims were full of eyes round about.

1:19 And when the living creatures went, the wheels went beside them; and when the living creatures rose from the earth, the wheels rose.

When Blumrich first read the accounts of Ezekiel, it was these descriptions of the wheels which caught his attention. Not only was the fact of these mysterious wheels in association with the overall vision intriguing, but their particular design was extremely important as well.

Ezekiel immediately recognized the basic form of the wheels as they were lowered from their storage in the bases of the helicopter units, but he also observed their unique feature of being somehow able to move in directions perpendicular to their normal lines of travel. What kinds of wheels were these? Blumrich recognized that the only way for wheels to go "on their sides" without turning normally was to have a main wheel composed of many smaller wheels which turned sideways to the direction of travel to the main wheel. Ezekiel accurately described the configuration as a "wheel within a wheel!"

To better understand this concept, it might be helpful to think of a string of beads formed into a perfect circle and set on end like a wheel. If the entire string could be rolled like a hula-hoop along the ground, it would be functioning like a wheel as it is normally used. But if the hoop were held stationary, and the individual beads touching the ground were turned, the hoop would not go forward, but sideways! Thus Blumrich had unlocked from Ezekiel's description a practical understanding of the unique wheel employed by the builders of this alien craft, and he has even applied for a patent for his own derivation of this design.

1:22 Over the heads of the living creatures there was the likeness of a firmament, shining like crystal, spread out above their heads.

1:23 And under the firmament their wings stretched out straight, one toward the other...

These two passages reveal the structure of the main body of the vehicle in relation to the helicopter units. To reiterate, the helicopter units were suspended on struts from four points on the outermost rim of the vehicle, rather like legs on a mushroom. At this position the units were partially overshadowed by the main body, which Ezekiel likens to a firmament. He carefully notes that it has the appearance of crystal with a definite shape which extends to just over the tops of the helicopter units.

1:25 And there came a voice from above the firmament over their heads;

1:26 And above the firmament over their heads there was a likeness of a throne, in appearance like sapphire; and seated above the likeness of a throne was a likeness as it were of a human form.

1:27 And upward from what had the appearance of his loins I saw as it were gleaming bronze, like the appearance of fire enclosed round about; and downward from what had the appearance of his loins I saw as it were the appearance of fire, and there was brightness round about him.

1:28 Like the appearance of the bow that is in the cloud on the day of rain, so was the appearance of the brightness round about. Such was the appearance of the likeness of the glory of the Lord. And when I saw it, I fell upon my face, and I heard the voice of the one speaking.

Finally, in Ezekiel's methodical scrutiny, he turns to the cockpit of the vehicle, where he deftly describes the control "throne" and the pilot, slightly obscured by reflections on the canopy and lights in the cockpit. The pilot appears to have been wearing a reflective bronze-colored flight suit, which may have added to the reflections in the cockpit.

If anything could have inspired an ancient observer to believe that it was a god, surely *this was it*. But even in his awed state, Ezekiel is careful to relate only what he sees, calling the pilot a "human form" and noting that someone, unknown to him, was speaking.

But the mystery is quickly shattered; it was the Lord of Israel, and that day Ezekiel was given an unenviable mission. To those who were left in Jerusalem after the first captivity, there was a strict warning. In case they felt that they were any better than those previously taken into exile, Ezekiel was to warn them that they too would be overthrown, and he was to make it clear to the people there that they had done no better at observing God's laws than the captives. Ezekiel was to become an intermediary between the Starmen and Israel:

Son of man, I have made you a watchman for the house of Israel; whenever you hear a word from my mouth, you shall give them warning from me. (Ezek. 3:17)

With his commission delivered, Ezekiel is removed from the place of contact and returned to his people via the alien vehicle, which Ezekiel calls the "glory of the Lord."

3:12 Then the spirit lifted me up, and as the glory of the Lord arose from its place, I heard behind me the sound of a great earthquake;

3:13 it was the sound of the wings of the living creatures as they touched one another, and the sound of the wheels beside them, that sounded like a great earthquake.

3:14 The spirit lifted me up and took me away, and I went in bitterness in the heat of my spirit, the hand of the Lord being strong upon me;

3:15 and I came to the exiles at Tel-abib, who dwelt by the river Chebar. And I sat there overwhelmed among them seven days.

The exact method of Ezekiel's conveyance is not made clear, but evidence suggests that he may not have flown inside the vehicle. Several clues indicate that he may have been literally plucked from the ground where he stood and transported in a quite traumatic, if efficient, manner.

When the vehicle arose from the ground, Ezekiel seems to say that the "spirit," as an invisible force, lifted him from where he stood and carried him through the air *outside* the ship as it flew. In support of this, Ezekiel's straightforward report makes no mention of his entering anything. Instead, as he is flying, he refers to the sounds of the wings and the wheels as they come from behind, and not from below him, as it would have been if he were riding in the cockpit.

His level of distress further supports this argument. It was not a smooth, quiet ride in an insulated cockpit, but a frightening, noisy experience made "in bitterness in the heat of his spirit!" Suspended by a force he could not see and flying at heights that only birds usually experience, he was in all probability a mere specimen undergoing a routine transfer from one point to another in the most expedient, efficient manner possible.

Other passages in the Bible allude to this force which physically transported people through the air:

Thou liftest me up on the wind, thou makest me ride on it, and thou tossest me about in the roar of the storm. (Job 30:22)

Elijah was lifted up and carried around in this manner several times before going up into heaven forever:

And as they still went on and talked, behold, a chariot of fire and horses of fire separated the two of them. And Elijah went up by a whirlwind into heaven. (II Kings 2:11)

It seems that others knew of this force which "caught up" Elijah for witnesses to this event who were across the river organized a search party for him:

And they said to [Elisha], "Behold now, there are with your servants fifty strong men; pray, let them go and seek your master; it may be that the Spirit of the Lord has caught him up and cast him upon some mountain or into some valley." (II Kings 2:16)

Phillip the evangelist had a similar experience after baptizing an Ethiopian eunuch near Jerusalem:

And when they came up out of the water, the Spirit of the Lord caught up Phillip; and the eunuch saw him no more, and went on his way rejoicing. But Phillip was found at Azotus, and passing on he preached the gospel to all the towns till he came to Caesarea. (Acts 8:39-40)

The possibilities for interpreting the terms "spirit" and "hand of the Lord" as used in Ezekiel obviously leave much room for speculation. By continuing to treat the Book of Ezekiel literally, we acknowledge the possibility that these terms as used may refer to a force for which we have no understanding. Was the force which Ezekiel called the "spirit" the same phenomenon which parted the Red Sea, tumbled the walls of

Jericho, and stopped the flow of the Jordan river? If so, it is a phenomenon of physics which we have yet to discover, and speculating further on something for which we have no understanding would be an exercise in futility.

In accepting the possiblity of the employment of an unknown force or forces by this alien culture, however, we do not in the least negate the basic argument for the evidence of extraterrestrial intervention in biblical literature. Indeed we would expect that an alien culture would possess some differences in understanding and employment of the laws of physics due to their independent evolution.

The key point to be made from our examination of this first "vision" is the overwhelming evidence of its physical nature. Despite the ambiguity of the term "spirit" used by Ezekiel, almost every other aspect of the account points to a true extraterrestrial encounter.

To recap, three factors attest to the authenticity of the event as a physical phenomenon: First, Ezekiel's lack of understanding throughout his first encounter, as well as his level of distress, is good evidence that he did not make up his experience. Second, through his lucid descriptions, we are able to see a vehicle of rather sophisticated design, according to a NASA engineer, with its attendant shapes and effects. Third, from this vehicle emerged a voice which gave Ezekiel a mission concerning his people, which suggests that the human form reported by Ezekiel was indeed a human pilot with a specific mission of his own. All aspects considered, this vision clearly demonstrates not only the rationality of the Starman Intervention Theory but the continuity between this and earlier references to the Lord of Israel.

But this contact, extraordinary as it was, was only the first in a series of four separate encounters recorded in Ezekiel. While the reports of subsequent contacts contain many details similar to the first, other parts of the material add significantly to our understanding of the Starmen and their mission and will be explored briefly.

The second encounter occurred sometime within the span of a year after the first and is given only the briefest attention with regards to the vehicle:

And the hand of the Lord was there upon me; and he said to me, "Arise and go forth into the plain, and there I will speak with you." So I arose and went forth into the plain; and, lo, the glory of the Lord stood there, like the glory which I had seen by the river Chebar; And I fell upon my face. But the Spirit entered into me and set me upon my feet; and he spoke with me... (Ezek. 3:22-24)

During this second visitation, Ezekiel is told of the impending doom of those remaining in Jerusalem and of their subsequent scattering, a message he must, in turn, relay to his people:

The word of the Lord came to me; "And you, O son of man, thus says the Lord God of the land of Israel; An end! The end has come upon the four corners of the land. Now the end is upon you, and I will let loose my anger upon you, and will judge you according to your ways; and I will punish you for all your abominations." (Ezek. 7:1-3)

Shortly thereafter, about one year from the first encounter, Ezekiel meets the Starmen for the third time. This was to be the last physical contact before the fall of Jerusalem, as the events which take place herein indicate what seems to be an evacuation of the temple and the removal of the Ark of the Covenant while Ezekiel looks on.

8:1 In the sixth year, in the sixth month, on the fifth day of the month, as I sat in my house, with the elders of Judah sitting before me, the hand of the Lord God fell there upon me.

8:2 Then I beheld, and lo, a form that had the appearance of a man; below what appeared to be his loins it was fire, and above his loins it was like the appearance of brightness, like gleaming bronze.

8:3 He put forth the form of a hand, and took me by a lock of my head; and the Spirit lifted me up between earth and heaven, and brought me in visions of God to Jerusalem, to the entrance of the gateway of the inner court that faces north...

8:4 And behold, the glory of the God of Israel was there, like the vision that I saw in the plain.

We confirm here that Ezekiel saw the same pilot figure, experienced the same force which lifted him, and saw the same vehicle as in the first encounter. The sequence of events indicates that the pilot came to Ezekiel's house to summon him in a smaller vehicle, however, and then flew him to Jerusalem "in visions of God'' (or literally through the air the way God would have seen the land while flying in his vehicle), where he saw the larger ''mother ship.''

Ezekiel is brought to the temple in Jerusalem, where he is told that Israel's transgressions are causing not only their own dilemma, but problems for the Starmen too, as hinted at here:

8:6 And [the leader] said to me, ''Son of man, do you see what they are doing, the great abominations that the house of Israel is committing here, to drive me far from my sanctuary?''

This passage seems to confirm our earlier speculation that the Starmen removed the Ark of the Covenant from Jerusalem prior to its fall and that it was not destroyed or hidden as others have speculated.

During this visit to the temple, several events occur which seem to have nothing to do with Ezekiel directly. It is apparent that there are other activities occurring around him involving other human-like beings so that at times the focus switches from Ezekiel to other matters in the courtyard. One such diversion occurs in chapter nine, which opens with the amplified voice of the leader calling forth seven men:

9:1 Then he cried in my ears with a loud voice, saying, ''Draw near your executioners of the city, each with his destroying weapon in his hand.''

9:2 And lo, six men came from the direction of the upper gate, which faces north, each man with his weapon for slaughter in his hand, and with them was a man clothed in linen, with a writing case at his side.

Note that Ezekiel sees nothing particularly strange or different about these men except that one is clothed in linen and the others are carrying ''destroying weapons.'' Physiologically, they are apparently like humans in every respect!

We must assume that the term ''destroying weapon'' connotes a device other than the standard sword or spear. The use of the words ''destroying'' and ''slaughter'' seem to imply that their destructive power was magnified, as if a grander scale of devastation could be wrought through them than through the standard armaments of Ezekiel's day.

Their first order was to go through Jerusalem destroying those who were committing the great atrocities against the Laws laid down by Moses. After they had finished the job, the man clothed in linen promptly reported back to the leader that the mission had been accomplished.

But it is the man clothed in linen whose actions are most interesting for his second task seems to be unconnected to the purpose for which Ezekiel was brought to the temple. Chapter Ten opens with a command from the leader for the specially clothed individual to approach the vehicle:

10:2 And [the leader] said to the man clothed in linen, "Go in among the wheels underneath the cherubim; fill your hands with burning coals from between the cherubim, and scatter them over the city."

10:6 And when he commanded the man clothed in linen, "take fire from between the wheels, from between the cherubim," he went in and stood beside a wheel.

10:7 And a cherub stretched forth his hand from between the cherubim to the fire that was between the cherubim and took some of it and put it into the hands of the man clothed in linen, who took it and went out.

This curious sequence of events demonstrates several significant things about the mission. For example, it is the man clothed in "linen" who is ordered between the wheels of the vehicle. He is instructed to remove "burning coals" or some substance requiring special handling, which suggests that he was wearing a garment designed to provide protection from heat or radiation. Protective clothing made of new "high-tech" fabrics is now routinely worn in the nuclear and chemical industries and by firefighters for protection against heat, radiation or chemicals.

This substance had to be removed from the area at the base of the main body of the vehicle, which Blumrich designated as the location of exhaust port for the ship's main engine. The actions of the linen garbed attendant reveal that the area was so "hot" that he could only approach to a point near the base of one of the helicopter units (which Ezekiel now calls cherubim).

At that point a mechanical arm at the side of one of the cherubs reached back and extracted the material from the hot area and handed it to the man clothed in linen, who then immediately departed. This use of the mechanical arm graphically demonstrates the rationale behind such a configuration. The pilot could operate the robot arm from his cockpit to handle the dangerous or noxious task without requiring him to leave the safety of the cabin, much the way our space crews operate their remote controlled arm from the cabin of the Space Shuttle.

This incident further reveals a vulnerability of the Starmen, their need for some type of protection and their human limitations. Even more interesting, it seems to point out a remarkable similarity between their solution to various limitations and our own solutions to very similar limitations!

But the question immediately arises: why was this incident included in Ezekiel's account? At one point Ezekiel was being warned of the impending disaster. Then suddenly he is left for a time to witness this seemingly unrelated event. The text says that the "burning coals" were to be spread throughout the city (Ezek. 10:2), but no reason is given for this act, and no damage is reported as a result. Futhermore, we know that the wicked and corrupt supposedly already had been eliminated by the other six "destroyers." It might therefore tentatively be concluded that what Ezekiel saw was some form of maintenance on the vehicle, such as the disposal of nuclear waste or similar material. The job could have been routine or an emergency which occurred while Ezekiel was being escorted. An emergency would explain why Ezekiel seems to have been left alone during this time.

Finally Ezekiel is brought back to the exiles for a third time:

11:22 Then the cherubim lifted up their wings, with the wheels beside them; and the glory of the God of Israel was over them.

11:23 And the glory of the Lord went up from the midst of the city, and stood upon the mountain which is at the east side of the city.

11:24 And the Spirit lifted me up and brought me in the vision by the Spirit of God into Chaldea, to the exiles. Then the vision that I had seen went up from me.

The impending disaster was obvious to the Starmen. Within five years of Ezekiel's first warnings, Jerusalem was destroyed. Spurred by a strong nationalistic spirit and a false sense of security, the remaining free Israelites rebelled against their Babylonian oppressors. But they had miscalculated, and Nebuchadnezzar was quick to lay siege to Jerusalem. In the struggles which followed, Zedekiah, appointed king over Judah by Nebuchadnezzar himself after the first exile, was killed, and the city fell in July 587 B.C. The temple and palaces were destroyed, and virtually all inhabitants were deported except for the poorest farmers and laborers.

Ezekiel during this time produced copious prophecies concerning Israel's doom. Then, some twenty years after the first encounter, he made his fourth and final recorded contact with the Starmen:

40:1 In the twenty-fifth year of our exile, at the beginning of the year, on the tenth day of the month, in the fourteenth year after the city was conquered, on that very day, the hand of the Lord was upon me,

40:2 and brought me in the visions of God into the land of Israel, and set me down on a very high mountain, on which was a structure like a city opposite me.

40:3 When he brought me there, behold, there was a man, whose appearance was like bronze, with a line of flax and a measuring reed in his hand; and he was standing in the gateway.

40:4 And the man said to me, "Son of man, look with your eyes, and hear with your ears, and set your mind upon all that I shall show you, for you were brought here in order that I might show it to you; declare all that you see to the house of Israel."

The last eight chapters of Ezekiel describe in minute detail the plan for the reconstruction of the temple and for the general organization of the Israelites after the exile. In order for Ezekiel to understand the plan, he is led around this city in the mountains to record its dimensions as given by his mysterious escort, who takes them directly from the city itself.

Let us examine the situation for a moment. It is extremely unlikely that the Starmen would have gone to the great expense and trouble of constructing an elaborate model of an entire city for Ezekiel's one-time instruction, and we are proceeding upon the assumption that these "visions" were more than simply dreams. Therefore, if we accept earlier passages, which plainly stated that God lived in the hills and mountains around Israel, especially Mt. Zion ("Who shall ascend the hill of the Lord? And who shall stand in its holy place?" Ps. 24:3; "His abode has been established in Salem, his dwelling place in Zion", Ps. 76:2), then we have no trouble in realizing that Ezekiel was indeed brought to one of the Earth bases of the Starmen!

The significance of this is that for the first time we have a direct eyewitness account of one of these bases, which up to this point had only been alluded to. The Garden of Eden, itself a Starman base upon a high mountain, was where it all began. But there was only a hint of the formal organization in the metaphoric expressions of the ancient Genesis author who told of the first hybrid beings who were evicted from the Garden and then prevented from returning by a robotic sentry. It is Ezekiel's account which verifies in retrospect our earlier hypotheses and is further testamony to the amazing continuity in the biblical accounts of this extraterrestrial intervention during ancient times.

Since the term "city" is used, and we must assume that Ezekiel knows the difference between "city" and "temple," we realize that Ezekiel describes only the portion of this lofty sanctuary which concerned his people, and that was the temple and its courtyard. Despite its limited scope, several details may be extracted for our use.

From the dimensions provided, it is obvious that this structure was much larger than the temple constructed by Solomon some four centuries earlier. The original sanctuary was built to house only the Ark and, though ornate, was only about the size of a small chapel. The structure under study by Ezekiel, however, was more like a fortress by design with perimeter walls approximately 284 yards, or nearly three football fields in length, and an area of more than 80,000 square yards!

The apparent reason for this larger dimension was that the new sanctuary was to serve as a landing pad for the alien vehicle. After receiving some instruction, Ezekiel was led by his escort to the east gate of the mysterious city to watch the arrival of such a ship which descends to the landing pad in the sanctuary with its landing lights on:

43:1 Afterward he brought me to the gate, the gate facing east.

43:2 And behold, the glory of the God of Israel came from the east; and the sound of his coming was like the sound of many waters; and the earth shone with his glory.

43:3 And the vision I saw was like the vision which I had seen when he came to destroy the city, and like the vision which I had seen by the river Chebar; and I fell upon my face.

43:4 As the glory of the Lord entered the temple by the gate facing east,

43:5 the Spirit of the Lord lifted me up, and brought me into the inner court; and behold, the glory of the Lord filled the temple.

After the landing of the vehicle, more instructions follow regarding such matters as proper conduct for priests and for the people, the division of the land, and the structure of the government after the exile. A particularly interesting segment deals with an entrance for the vehicle in the future sanctuary:

44:1 Then he brought me back to the outer gate of the sanctuary which faces east; and it was shut.

44:2 And [the Lord] said to me, "This gate shall remain shut; it shall not be opened, and no one shall enter by it; for the Lord, the God of Israel, has entered by it; therefore it shall remain shut."

In Ezekiel 43:4 we read that "the glory of the Lord entered the temple by the east gate" in a very physical way, and Ezekiel is brought into the inner sanctuary, presumably to witness how this was done. Thus there can be no doubt at this point that the sanctuary was specifically designed as a secure landing point. As stated by the Lord himself, "Son of man, this is the place of my throne and the place of the soles of my feet." (Ezek. 43:7)

The location of the sanctuary proposed for the Israelites after the exile also indicates that the Starmen intended a new and expanded role for the priests, who would operate in a closer collaboration with the Starmen in a more secure environment. C.H. Patterson capsulizes:

The temple is to be built outside the main part of the city and constructed in a manner that will make it possible to keep out all of those persons and objects which might contaminate the holy place in which Yahweh will dwell in their midst. A sharp distinction is introduced between priests and Levites in order that only qualified persons should enter the building even for the purpose of keeping it clean. The highest official will no longer be the king, but rather the high priest, thus indicating that political affairs shall always be made subordinate to religious considerations. (144)

In an apparent throwback to the days of old, the plan was to reinstitute old ritual to allow Starmen to again return to the people in glorious and awe-inspiring displays, but at the same time protect them from the inherent dangers involved in the close relations.

Significant passages deal with just such protective measures, as manifest in regulations for conduct in the temple compound:

44:5 and mark well those who may be admitted to the temple and all those who are to be excluded from the sanctuary.

44:17 When they enter the gates of the inner court, they shall wear linen garments; they shall have nothing of wool on them, while they minister at the gates of the inner court, and within.

44:18 They shall have linen turbans upon their heads, and linen breeches upon their loins; they shall not gird themselves with anything that causes sweat.

44:19 And when they go out into the outer court to the people, they shall put off their garments in which they have been ministering, and lay them in the holy chambers; and they shall put on other garments, lest they communicate holiness to the people with their garments.

44:20 They shall not shave their heads or let their locks grow long; they shall only trim the hair of their heads.

44:21 No priest shall drink wine, when he enters the inner court.

44:23 They shall teach my people the difference between the holy and the common...

Without much imagination, it is easy to believe that the priests were actually taking precautions to protect their health from the effects of radiation, static electricity or some similar hazard. It is extremely interesting that the clothing specified for ministering to God had to be removed before the priests left the inner court in order to prevent other people from being contaminated with "holiness!" Later editors were understandably distressed with this vision which specified rules for preventing the communication of holiness, but only because they had no way of knowing that objects with such a quality had potential negative effects upon the health of those who did not take the necessary precautions. And it must be reiterated that the Starmen took great pains to prescribe specific, detailed systems for ensuring safety. It was only when set ritual was circumvented or carelessly applied that innocent lives were lost.

The other ordinances—such as keeping the hair trimmed, refraining from drinking alcoholic beverages prior to ministering, and limiting access to the inner court—were just common sense precautions and are quite compatible with our hypothesis that the ritual served a pragmatic split-level function of allowing primitives to administer to the Starmen while providing maximum protection for the masses.

Interestingly enough, this elaborate temple was never built, nor did the Starmen ever return to "dwell in the midst of the people" of Israel after the exile ended. Exactly why the Starmen chose to abandon their original plan to create a race of "priests" on Earth is not clear. What is clear is that events recorded in the *Book of Ezekiel* mark a fundamental change in the character of the relationship between the Starmen and the Hebrews. What was at first expressed by the Starmen as a sense of loss and dismay over the waning Hebrew conformity subsequently degenerated into bitter condemnation:

> *"Wherefore as I live," says the Lord God, "surely, because you have defiled my sanctuary with all your detestable things and with all your abominations, therefore I will cut you down; my eye will not spare, (*another reading is, *'I will withdraw') and I will have no pity."* (Ezek. 5:11)

In hindsight, this deterioration, though obviously much more complex than this, was at least partly the result of the tremendous combined socio-political, cultural, ecomonic and human forces at work in the Hebrew environment of the period which seriously eroded the once solid allegiance to the Starmen. Surrounded by superpowers and strategically located at the crossroads of a great trade network built by their predecessors, the Hebrews found themselves pushed and pulled in a multitude of directions at any one time. Without the powerful, charismatic leadership of King David or King Solomon to maintain a tenuous alliance among the tribes, Israel became as a wounded fish in a sea of sharks.

If the Starmen did indeed come and go at various intervals throughout Israel's history as proposed earlier, it is clear that they were not present when Solomon died. So it was that without immediate and decisive action to install a successor, the external forces quickly polarized the once unified nation, which then split into rival states. When the Starmen returned, it was too late, for the momentum of the tide had built quickly and, despite the direct word from the Starmen through prophets such as Ezekiel, the wounded fish had already been divided for the kill. To physically intervene on behalf of the Hebrews at this point must have seemed a useless alternative to the Starmen, for the corrupt behavior of the Hebrews was a deeply rooted problem, and intervention would have meant only a temporary reprieve from what was already a mortal wound.

The Starmen's solution to this dilemma was Jesus. Manifest in this final "Son of God" was a salvage plan based upon an entirely different set of assumptions and criteria which were more realistic and more attuned to the new political climate and to human nature. Ezekiel thus provides us with the important transition between the Intermediate Order and the New Order, the final phase in the evolving Starman plan.

Jesus

Few other individuals, if any, have had so great an impact upon history as Jesus. A lifetime of study would, in all probability, yield only an incomplete understanding of his full impact upon mankind. This being the case, we must at least try to understand, if only at a very superficial level, why Jesus was important to the Hebrews under the Starman Intervention Theory and then why his teachings became important to all mankind after his death.

The Gospel stories of Jesus are so well known that it is not necessary to recount every episode here. There are, however, key events of his life which are significant in the light of our theory. For example, Jesus' ties to the Starmen are clearly demonstrated in such events as his "virgin" birth, the moving "star" of Bethlehem, the "spirit" which descended over Jesus at his baptism, the transfiguration, and the ascension. The continuity of events between the Old Order and the New Order is further supported by the essential elements of this alien presence during Jesus' time: vehicles described as bright lights or clouds, Starmen as angels, unseen voices, and the "pure vine" metaphor, continue to be found in the New Testament.

Much has been written about Jesus and the birth of Christianity. While Christians traditionally think of Jesus as the son of a spiritual God, some sources find him to have been merely a great man. A scholarly work by H.J. Schonfield entitled *The Passover Plot,* for example, proposes the theory that Jesus orchestrated his own crucifixion simply because he believed that he was the Messiah. Interestingly, both of these approaches to understanding Jesus have some merit under the Starman Intervention Theory. By analyzing and combining both secular historical information and the scriptural information from the Gospels, we find that Jesus really could have been a "son" of the "living God" who passed through trials and tribulations imposed upon him by the unforgiving religious and political climate of his time.

But in order to understand the phenomenon of Jesus, certain crucial elements of his life and times must be put into their proper perspective. By digressing for a moment to touch upon some of these crucial elements from a secular standpoint, we may gain some valuable insight into Jesus, into his impact upon the world, and into the hypothesis that Jesus was, in fact, the last of a long lineage of alien hybrids sent to the Earth to create a race of "priests."

First we must understand that the Gospels are our primary source of information on Jesus. Little more than passing references to Jesus are found in secular writings by such historians as Josephus, Tacitus, Suetonius and Pliny the Younger, (145) so that our information about him remains crippled with bias and mystery.

Second, despite their impact of incalculable proportions, the Gospels came into being in a way that few seem to understand. Many do not realize, for example, that of the authors of the Gospels (Matthew, Mark, Luke and John), *not one* saw or knew Jesus, and all wrote from oral traditions and accounts which were circulating between sixty and ninety years after Christ's death! Furthermore, of the twelve apostles- Andrew, the two Simons, John, the two Jameses, Phillip, Jude, Bartholomew, Matthew, Thomas and Judas - not one was an author of Gospel material. Two of the Gospels use the names of, but were not written by, Matthew and John. This appears a common misconception even among contemporary Christians. One might ask then, "Who did write the Gospels?"

Briefly, the Gospel of John is thought to have been written between A.D. 95 and A.D. 105 by a single author who was also named John (John the Elder, as distinguished from John the

Apostle), using oral traditions and possibly an eyewitness as his sources.

The Gospel of Luke is thought to have been written by a man of the same name who, according to *Harper's Bible Dictionary,* was a Greek physician and a traveling companion of St. Paul. He was also the likely author of some of the book of Acts, though he is never mentioned by name in either of these writings. Luke took on its final form probably about 75 A.D.

The Gospel of Matthew was written by an unknown author, thought to have been a Christian Jew, who assembled various written and oral Greek traditions circulating about the same time as those of Luke, but the varying viewpoints indicated in the scriptures suggest that this Gospel was assembled in a location with both Christian and Jewish influences, such as Antioch in Syria. (146)

Considered the sourcebook for the other three Gospels, Mark is the oldest of the four and has influenced the authors of the other three Gospels to some degree, especially those of Matthew and Luke. It can be dated between 64 and 85 A.D., (147) and was written probably by Mark himself, who drew from recollections of his work as an interpreter for a later apostle named Peter. Peter and Paul were supposedly martyred in Rome about 64 A.D., which prompted Mark to record the accounts which Peter had revealed to him, and this Gospel may in fact have been composed from the memoirs of Peter, now lost. (148)

A question which arises at this point is, "Why did these various disciples wait so long to write about Jesus' life and deeds?" The answer has basically two parts. To begin with, Jesus taught that his kingdom was "at hand," and therefore no one saw any need to record anything during Jesus' time. When months and years passed without the arrival of his kingdom, however, it was decided to record the events in order to preserve them while the faithful waited for what they felt was still to occur, even if it was delayed.

Additionally, as time passed, the life and deeds of Jesus were increasingly glorified by adherents to his teachings. As the early movement became the church, its momentum increased manifold, and people who had never even known Jesus were suddenly inspired to write about him in testimony to their conversion and as affirmation of their faith in his teachings.

Consider the condition of the pre-Christian Hebrews at this point. They had centuries earlier rejected or abandoned the guidance of the Starmen and the counsel of their proxy leaders, who sought to mold them into a race of Earth administrators for a future world order. But the "pure vine of Israel" had so corrupted itself, both biologically and religiously, that they were about to forfeit their privileged positions as "hosts" to the Starmen of the universal order.

By the end of the first century B.C., the Hebrews had lost all touch with the concept of a "living" human God. Their eventual accommodation of the Greek traditions of Hellenism and various other pagan religious adulterations brought by conquering armies, coupled with the absence of any material Starman activity for many years, caused the physical being once known as a god to become an abstraction, a "spirit" to his own people.

Moreover the political climate of the Hebrews around the time of Christ was one of repression and persecution. The conditions among the Hebrews were ripe for a savior who would come in all of his historical might and glory to once again unite the Hebrews and to shed the oppression of Rome, as had been done so long ago in Egypt against Pharaoh, thereby establishing the kingdom of God on Earth forever.

But this longing for a savior was no random event. Hebrew scholars had long been concerned with interpreting Bible passages relating to God's treatment of Israel, and it was determined that their period of repression was a punishment of limited duration.

It was widely believed that chapter nine of Daniel indicated 490 years of punishment:

"Seventy weeks of years are decreed concerning your people and your holy city, to finish the transgression, to put an end to sin, and to atone for iniquity, to bring everlasting righteousness, to seal both vision and prophet, and to anoint a most holy place. Know therefore and understand that from the going forth of the word to restore and build Jerusalem to the coming of the anointed one, a prince, there shall be seven weeks." (Dan. 9:24-27)

Therefore, after the commandment by Cyrus the Great to rebuild Jerusalem some time between 539 and 536 B.C., the prophecy was considered begun. A fervor of unparalleled proportions thus began around 46 B.C., according to Schonfield (149), when a phenomenon that became known as the *Messianic Hope* or the expectation of the coming of a great messiah who would restore Israel to its former glory began to grip the Hebrew people. Although many complex factors contributed to this

Messianic Hope, there are three major circumstances, according to Schonfield, which shaped the fervor of those times.

The first of these conditions was the change of attitude toward the Bible which was then in existence. Basically the Hebrew Bible had developed to that point in three stages;

The Hebrew Bible consisted of three divisions: The Law, The Prophets (Joshua to Malachi), and The Writings (beginning with Psalms and including the book of Daniel). The divisions represent stages of acceptance into canonicity. The Law, consisting of the five books of Moses, had binding force by the fifth century B.C. or not much later. The Prophets did not acquire their force until about the third century B.C. The Psalms and some other books soon formed the basis of the third division, which was finally settled at the end of the first century A.D. (150)

The Hebrews held their scriptures in the utmost reverence and authority. They were the word of God and, as such, were binding upon all Hebrews. Once these writings became recognized as the word of God, the weight of their messages was greatly magnified. This was particularly true of the canonization of The Prophets, which detailed the coming of the Messiah.

The second condition was the dangerous drift away from the strict avoidance of other religions and flirtations with Hellenism, which was having an increasing impact on both Hebrew thought and way of life. This was seen by some factions as a corruption of the "pure vine" concept and would be severely dealt with by God at the end of the age.

Together the prophecies of the Hebrews' literature combined with the political and religious oppression they suffered under such rulers as Antiochus Epiphanes (175-62 B.C.) and Augustus Herod, (Herod the Great; 37-4 B.C.) served to heighten Hebrew consciousness of the significance of the period. The writings were scanned for every possible indication of the prophecy. Every manner of interpretation and every shade of meaning was applied to the scriptures. A full spectrum of approaches was applied, from liberal to conservative, eclectic to pure. Various modes of the Hebrew religion thus evolved with the intention of capturing the title of "righteous remnant of Israel." Schisms among the Hebrews widened as debates intensified over such issues as "proper" scriptural interpretation, how a Hebrew should live to be "righteous" in the eyes of God, and over the character and manifestation of the expected "Anointed One."

With all of the various social and political pressures on the Hebrews during those times, it is reasonable to expect that they would eventually splinter into several religious and political factions as they sought ways to cope with their changing environment. There were perhaps four or five small but significant Hebrew religious movements or sects worth mentioning here which had a great impact upon the events and upon Hebrew thought during the period. They included the Sadducees, the Pharisees, the Essenes, the Zealots and the Gnostics.

The Sadducees were the oldest of the major factions. Generally well-educated, wealthy and conservative, these Hebrews, though comparatively few in number, exercised some degree of influence with the occupation government of Rome. The priesthood rested with the Sadducees, and their strict adherence to the Laws of Moses was their overriding preoccupation. They tolerated little flexibility in their scriptural interpretations and paid little heed to oral traditions and unwritten laws circulating outside Mosiac Law.

They were ambivalent about their treatment of gentile religious precepts however. On one hand they rejected the ideas of the resurrection of the dead, spirits or angels. On the other hand, they encouraged the examination of ideas from other religions in order to appreciate any potential contributions they might make.

The Pharisees originated as a separatist movement shortly after the emergence of the Sadducees, sometime around the first century B.C. Their attitudes were varied, and they formed the broad range of the Hebrew religious center, being roughly divided into three main groups according to C.F. Potter:

On the Pharisee right were the followers of the comparatively conservative Shammai; in the middle were the Hillellites whose grandson Gamaliel was Paul's teacher in Phariseeism. Most deviating of the Pharisees were the Apocalyptic Pharisees, a "modernist" group who laid much stress on such books as Daniel, Ezekiel and the non-canonical Enoch. (151)

(As a note, Shammai, Hillel and Gamaliel are considered three of the most significant Jewish thinkers of all time by many scholars.)

The Pharisees formed by far the largest of the sects, and they also enjoyed a degree of political clout with the civil government. This group, unlike the Sadducees, did embrace varying shades of the unwritten laws and oral traditions which they later canonized as "The Prophets" (200 B.C.), and "The Writings" (90 A.D.) and was the prime stimulus for the expansion of the scriptures during the period. They speculated on the concepts of "spirits" and "demons," the idea of the dual control of man both through his own actions and through divine guidance, the idea of the resurrection of the body, and generally a more liberal interpretation of the Laws of Moses as they applied to the circumstances of everyday life. They did not bind themselves to the letter of the Law, but rather considered how the Law might be applied to living and understanding the times through it. Together the Sadducees and the Pharisees, though rival factions at the time, were in agreement that Jesus was a heretic, and it was through their combined efforts that he was so branded and executed.

The third and perhaps the most significant of the sects from the standpoint of their contribution to our understanding of the inter-testamental times (the tumultuous period between the writing of the Old and New Testaments of the Bible), are the Essenes of Qumran. Little was known about this sect until the discovery in the mid-forties of the famous Dead Sea Scrolls, which are thought to be some of the oldest existing source documents of Bible literature.

The first scrolls were discovered in 1945 by a Bedouin shepherd who found some leather parchments in a wilderness cave about eight miles south of the ancient city of Jericho. Not realizing what he had found, he proceeded to use the better portions for sandal straps! The remainder of the scrolls hung in his home for more than two years until a friend persuaded him to take them to an antique dealer for appraisal, or so the story goes. For several years thereafter, tribesmen scoured the cliffs and surrounding countryside in search of the precious documents, which were then ransomed to harried archaeologists for exorbitant fees. (152)

According to ancient historians Josephus and Philo, there once existed about 4,000 Essenes in all with a significant portion of those residing in Qumran. We know that Qumran was an Essene library and scriptorium where members spent much of their time studying the scriptures, transcribing existing books and even producing literature of their own until their camp was destroyed by the Roman Tenth Army in the summer of 68 A.D.

The Essenes retired to the desert wilderness between 145 and 130 B.C., according to *Harper's Bible Dictionary* (153), to escape the deplorable secularization and corruption which they saw occurring to their religion among their Hebrew brethren. In their isolated environment, they lived in a sparse communal arrangement and became totally absorbed with the Law. Through their studies, they perceived that the end of an evil age was near, and they made preparations for the coming of the Messiah and the establishment of a new kingdom on Earth.

Several authorities, including Potter and Schonfield, contend that the teachings of Christ very closely approximate those of the Essenes in many respects and assert that at least part of Jesus' youth may have been spent training in these Essene camps.

In the Bible no mention is made of these "saints," as they were sometimes called, but Jesus several times preached against the Sadducees and the Pharisees in the Essene tradition, saying that they taught the Laws of Moses but lived hypocritical lives:

The Scribes and the Pharisees sit on Moses' seat; (teach his Law), so practice and observe whatever they tell you, but not what they do, for they preach but do not practice. (Matt. 23:1)

Take heed and beware of the leaven of the Pharisees and the Sadducees... (from Jesus' parable against the teachings of the two sects; see Matt. 16:6-12)

But the scrolls themselves are perhaps the best indicators of the relationship between Jesus and the Essenes:

The translations of the Qumran scrolls show quite unequivocably that vital parts of the Gospels originate from the Essene school, that Jesus' style and way of life followed the customs of the Essenes, and that parables, such as Jesus used - indeed, whole sermons attributed to him - had been taught by the Essenes long before him. (154)

The fact that we do not find more Essene literature in our Bibles, according to Potter, is due to the political infighting among the major Hebrew sects and the early gentile Christians:

The Essenes, you see, were the intermediate step between Judaism and Christianity, and when the Jews and Christians came to hate each other, due largely to the charge by Gentile Christians that the Jews were "Christ-killers," both sides turned on the Essenes and eliminated the Essene religious literature from their canonical scriptures as far as possible, or prevented its inclusion. (155)

With the discovery and interpretation of the Dead Sea Scrolls, we are just beginning to understand the full significance of this once obscure sect and its role in shaping the teachings of Jesus and the birth of Christianity. Potter even goes so far as to predict that evaluations of the scrolls will lead to a rewriting and reorganization of the Bible (156) to correct later omissions and adulterations to the Bible text! Whatever the result, the Essenes are destined to be a major source of controversy in the coming decades.

A fourth faction was the Zealots. These were groups of Hebrew patriots, rather loosely confederated, who believed that the way to rid themselves of the Roman yoke was by force:

These were the revolutionary messianists who believed in the use of violent methods in order to gain freedom from their oppressors. They were feared by the Romans because of their tendency to stir up rebellion against the recognized government. (157)

As much a political coalition as a religious sect, these members formed what Josephus once called "the fourth philosophy" (158), and it is known from Acts 1:13 that Christ chose Simon the Zealot as one of his twelve disciples. Others have argued that Judas Iscariot and, indeed, Jesus himself were Zealots as well.

Other groups which are known to have contributed to the Messianic Hope movement in various ways which are beyond the scope of this book to describe in detail are the Zadokites, a group of reformed priests who resented the way the Sadducees had made political offices out of the priesthood (159), the Amhaaretz ("people of the land," or the common Hebrews), and the Gnostics, who were not necessarily a separate and distinct group either but a loose confederation whose members espoused a philosophy which combined mystic elements of eastern religions with Judaism to produce the concept of salvation through the revelation of a "savior," such as Christ.

* * *

It should be quite evident at this point that the Hebrews were fully cognizant of their condition and of the significance of the period into which they had entered. Although most fully expected a great messiah, they were united on little else. Their once coherent, integrated nation was but a memory by this time, and each faction realized that as the time approached, there could only be one "righteous remnant of Israel." The stage was now set for the arrival of the Messiah.

Let us briefly reiterate three alternative approaches to understanding the events in Jesus' life. The first approach is to treat the biblical narratives in the traditional manner, accepting —as most Christians do—that extraordinary events were wrought by a spiritual God through Jesus as "miracles." The second approach that can be taken is to try to substantiate all actions by Jesus in terms of secular, terrestrial factors and Jesus' own acts of self-determination, under the Schonfield model.

H.J. Schonfield hypothesizes that Jesus orchestrated his own ministry to fulfill biblical prophecy and that it was through these prophecies that his actions were designed and scheduled. Schonfield's scholarly and controversial theory that Jesus was merely a product of his own times also postulates that Jesus skillfully planned, with the help of an unnamed cohort, to suffer for a time on the cross, be removed to a secret tomb and "miraculously" arise from the "dead" some time later to claim resurrection. While it is not possible to present Schonfield's elaborate evidence here, it should be possible, without much imagination, to see that the theory is not entirely preposterous. More ridiculous ideas have certainly been proposed!

However, while Schonfield's insights into the social, political and religious climates of Jesus' time seem quite tenable, as do his assessments of Jesus' conscious orchestration of his ministry and resurrection, his conclusion that Jesus was merely a great man and a brilliant operator intrinsically denies the extraordinary events which the Gospels record as truthful testimony to Jesus' qualifications as the "Son of God."

Understandably Schonfield, like other secular harmonizers, must invariably dismiss all extraordinary events in terms of later glorifications, myths and legends if they cannot be explained in terms of rational scientific reasoning or logic. This leads us to the third alternative approach, which seems to find its own unique middle ground between Schonfield's complete denial of all extraordinary events surrounding Jesus and the complete acceptance of all extraordinary events portrayed in the New Testament as manifestations of a spiritual God incarnate.

Under such a presumption, all extraordinary events reported can be explained either in terms of physical phenomena or as myths, legends and later glorifications of Jesus. As stated before, the key to understanding these phenomena is the time factor. Some of the phenomena identified in the New Testament clearly have a resemblance to physical phenomena and concepts which we know intimately and take for granted today, such as aircraft, space travel, artificial illumination, electronic voice amplification and biological manipulation. Where secular harmonizers begin to grope for answers as they try to explain these phenomena only in terms of what we believed about a technologically primitive culture in ancient times can be logically and rationally explained by introducing the notion of an advanced alien culture visiting our primitive Hebrews!

Some extraordinary events, however, are readily identified as myths, borrowed legends or later glorifications through an anaylsis of history and literature. Thus, like Schonfield and others, we can safely deny, or at least defer, judgment on such events as Jesus' healing the leper (Matt. 8:1-4), Jesus' turning water into wine (John 2:1-11), or of his miraculous multiplying of the loaves of bread and fish to feed the multitude (Luke 9:12-17).

Therefore the Starman Intervention Theory provides a version of events which specifies that Jesus was conceived after a visit by an angel (Starman) who employed a method of biological intervention already detailed, that his life and especially his ministry were part of a larger plan by the Lord (Starman leadership) to retrieve a wayward Hebrew kin, and that after the crucifixion, Jesus was removed from the cross, placed in a tomb and then removed into Heaven (space) by the Starmen in accordance with the theory that these were all physical events made supernatural by time and lack of understanding. Let us examine the scriptural testimony to these events in the light of the Starman Intervention Theory.

And so it was the Starmen returned in keeping with scriptural promise and prepared to deliver to the Hebrews a final hybrid proxy-leader in the tradition of the "pure vine" of Israel. Jesus was to be the last in a long lineage of "test-tube babies" born unto a virgin "host" through artificial insemination and who would be trained and counseled to carry out the final effort to collect the "lost sheep of Israel" and establish the priestly kingdom on Earth.

The first part of this plan involved finding suitable female candidates to carry two hybrid infants. The first child to be "planted," we are told in Luke, was John the Baptist, and the second was Jesus. Recall from an earlier section that two kin from the line of David were chosen for this task. The virgin

Mary was to bear Jesus, and her cousin Elizabeth (160) would give birth to John.

First the angel Gabriel was dispatched to Elizabeth to carry out the "planting of the seed" and to counsel both Elizabeth and her husband, Zechariah, as to the purpose of the mission. They were told that despite Elizabeth's "barrenness," her womb would be "opened" as in the days of old and the child born to her would have a great significance for Israel.

Six months later, Gabriel came to Mary and told her of her selection by God to carry Jesus. She too was instructed that despite her virginity, she would conceive and bear a son through the Holy Spirit, which would "come upon her," and the power of the Most High would "overshadow her" to create the child who was to become the awaited Messiah. (Lk. 1:35)

As the time of Jesus' birth approached, the Starmen became noticeably more visible. When Jesus was finally delivered in Bethlehem, the Starmen sought to publicize the event among the people in order that they would know that the Messiah had now arrived. One such event is clearly demonstrated in Luke:

Chapter Two

2:8 And in that region there were shepherds out in the field, keeping watch over their flock by night.

2:9 And an angel of the Lord appeared to them and the glory of the Lord shone around them, and they were filled with fear.

2:10 And the angel said unto them, "Be not afraid; for behold, I bring you good news of a great joy which will come to all the people;

2:11 for to you is born this day in the City of David a Savior, who is Christ the Lord."

2:13 And suddenly there was with the angel a multitude of the heavenly host...

2:15 When the angels went away from them into heaven, the shepherds said to one another, "Let us go over to Bethlehem and see this thing that has happened, which the Lord has made known to us."

The immediate reaction to this brief report is that it has a remarkable degree of similarity with earlier recorded Old Testament encounters. Here, as in other accounts, something descended upon the witnesses which was so ominous that it "filled them with fear." They could make out no particular form because the "glory" or brightness of this phenomenon "shone all around them," blinding them from direct observation.

From the information provided in Luke, at least two different conclusions can be drawn about this phenomenon: Traditional interpretations assume that this was a miraculous appearance by spirit angels. But in our times, the more rational alternative from a purely physical standpoint is that the shepherds were contacted by the Starmen from a vehicle hovering overhead. Consider, for example, that in order for a pilot of a modern night flying helicopter to locate a small group of people, he would, upon approaching the ground, switch on his powerful floodlight or landing light to be assured of clearing tall trees and other obstacles as well as to facilitate spotting the people.

The encounter was brief and businesslike. From this vehicle, obscured by the blinding light, came a voice which told the shepherds not to be afraid. Note that the voice of the "angel," or ship, could just as easily have been from an external speaker, such as the "bullhorn" used on modern police helicopters to give instructions to people on the ground. As the shepherds were being told of the birth of the Messiah, the first ship was joined by a "multitude of heavenly host." The spectacle made a deep impression upon the simple shepherds about the importance of the child, prompting them to depart "in haste" to see this new babe.

Some who witnessed the vehicle from a distance may have confused it with a bright star, except that this mysterious star moved through the heavens. The three wise kings are said to have followed such a "star" until it came to rest over the site of the delivery:

They went their way; and lo, the star which they had seen in the East went before them, till it came to rest over the place where the child was. (Matt. 2:9)

The "star" in the wise men story has been theorized by some to have been merely a comet or a rare conjunction of two planets such as Saturn or Jupiter in the night sky (161), but such speculations do not explain how the light "came to rest" directly over the spot where the child was delivered. When the light stopped, the three wise men were able to go directly in to see the child! One has only to accept that the light was, in fact, from an alien vehicle to be able to read the passage literally.

The next significant event in the life of Jesus was his baptism and his initiation into his ministry. Prior to this event, little is known of Jesus' childhood and early manhood. It is said in Luke that he was a child of exceptional intelligence and understanding who astonished the teachers in the temple of Jerusalem (Lk. 2:46-48). As for John the Baptist, even less is known. Luke 1:80 tells us simply that after his birth, John "grew and became strong in spirit, and he was in the wilderness till the day of his manifestation to Israel."

The symbolic and climactic initiation of Jesus into his ministry was to be accomplished by an historic meeting of these two men, in which John would baptize Jesus on the shores of the River Jordan. Prior to this event, John had been called by God from his wilderness seclusion to begin spreading the word around Judea of the coming of the Messiah. Soon he became well known for his fiery messages, and people would come to learn of his messianic prophecies and to be baptized in the river.

Finally it happened that Jesus came, and he too was baptized by John in the sight of a "multitude" of people. But the most important aspect of this event occurred after the baptism:

> *Now when all of the people were baptized, and when Jesus also had been baptized and was praying, the heaven was opened, and the Holy Spirit descended upon him in bodily form as a dove, and a voice came from heaven, "Thou art my beloved Son, with thee I am well pleased."* (Lk. 3:21-22)

In a dramatic display, with a multitude of people looking on, the Starmen descended over Jesus "in bodily form" in their vehicle and introduced Christ, as they had done for the shepherds at the time of Jesus' birth. There can be no mistake about this; it was a calculated act directed at a specific audience which was gathered by John. It was intended that this display, like those for Moses, Joshua and others, would impress upon the people the weight of the person who was being introduced. He was their new proxy leader.

Once Jesus had been introduced and the people could see that the Messiah had come, we are told that the "Spirit" immediately "drove" him out into the wilderness where he was both tempted by Satan and ministered to by angels for forty days (Mk. 1;12-13). Despite the addition of the temptation by Satan, the forty-day period in which Jesus was also counseled by "angels" corresponds to the forty days Moses spent on Mt. Sinai while he received instructions concerning his mission for Israel.

Whether Jesus fully understood the purpose and his destiny up to this point is not made clear, but there can be no doubt that from the baptism forward, he acted as one with a very strong sense of purpose:

...his calculated moves, his symbolic actions such as the forty days in the wilderness and the choice of twelve apostles, his staging of the triumphal entry into Jerusalem and the Last Supper all testify to his dramatic consciousness, as do many of his gestures and declamations. (162)

Jesus' mission involved operating in a very difficult environment. On one hand he was attempting to reach out to his "lost sheep" to form the base of the new kingdom. On the other hand, he had to negotiate an adverse religious and political situation which could at any moment cut short his vital ministry. This concern was underlined by his actions. Shortly before his baptism, and for a time afterward, Jesus healed, drove out demons and preached to the people with the stipulation that his identity and purpose not be revealed until the specified time. There was a strict timetable to observe in order to have all events fall into place, and he could not afford to be revealed before all was ready:

And he strictly charged them that no one should know this... (Mk. 5:43)

And [Jesus] sternly charged him, and sent him away at once, and said to him, "See that you say nothing to anyone; but go, show yourself to the priest, and offer for your cleansing what Moses commanded, for a proof to the people." (Mk 1:43-44)

In order to more effectively reach the masses, Jesus enlisted the help of twelve apostles, whom he chose from among the common people. According to the Gospels, these intimates were given a very special place in the new kingdom which was to come:

You [apostles] are those who have continued with me in my trials; as my Father appointed a kingdom for me, so do I appoint for you that you may eat and drink at my table in my kingdom, and sit on thrones judging the twelve tribes of Israel." (Lk. 22:28-30)

Although it is clear that these apostles did not understand everything which Jesus did or taught, they were privileged with certain knowledge of the mission not available to others:

The disciples came and said to [Jesus], "Why do you speak to them in parables?" And he answered them, "To you it has been given to know the secrets of the kingdom of heaven, but to them it has not been given." (Matt. 13:10)

Both Jesus and his apostles had to work in a sensitive political and religious climate, it is true. But Jesus also had the added dimension of being a hybrid who on occasion interacted with the Starmen. In speaking of these "secrets of the kingdom of heaven," Jesus was probably speaking of conditions in his father's world, in "heaven," the place on which the new kingdom on Earth was to be patterned in his near future. Therefore Jesus spoke in parables, not only for political camouflage, but so that his teachings might also pass uncluttered to his disciples, who would have had no understanding of such concepts. We may now examine some of Jesus' teachings to substantiate this premise.

The first important thing taught by Jesus and which tied him to this "kingdom of heaven" was that he was sprung from his people in the tradition of the Old Order. Said he, "I am the true vine, and my Father is the husbandman." (John 15:1-2) Thus, by his own admission, Jesus tells us that he was a new "branch" raised up by the Starmen, confirming our hypothesis of the injection of a hybrid proxy leader under the "pure vine" metaphor.

A second important teaching was that Jesus was not God himself, but the Son of God:

I do nothing on my own authority but speak thus as the Father taught me. (John 8:28)

For I have come down from heaven, not to do my will, but the will of him who sent me. (John 6:38)

As mentioned earlier, the Bible teaches repeatedly that God was a *living* entity:

But the Lord is a true God; he is the living God and the everlasting King. (Jer. 10:10)

For they themselves report concerning us what a welcome we had among you, and how you turned to God from idols, to serve a living and true God... (I Thes. 1:9)

In fact Jesus himself taught that he was the Son of the living God. Simon Peter, when asked by Jesus who he thought Jesus really was, replied, "You are the Christ, the Son of the living God." (Matt. 16:16) However twisted the term "living God" has become through the ages due to factors already discussed, it cannot be denied that God was once literally thought of as a living being. By Jesus' time, the term seems to have already begun its metamorphosis into a spiritual sense with its prior physical interpretation growing ever more difficult to believe for many Hebrews, who had not known God's presence in so long. When Simon Peter acknowledged Jesus' relationship to God, he was blessed and complimented for his insight and trust in Jesus' word even though he may not have understood exactly how Jesus meant "living God."

Beyond scriptural proof of Jesus' literal son-ship to a living God, it must also be emphasized that the deity of Christ was actually established years after Jesus' resurrection by a group of clergy and the emperor Constantine at the First Ecumenical Council at Nicaea, which convened in the year 325 A.D. The overriding purpose of the council of some 318 bishops, according to Von Daniken, was political:

...[Constantine] wanted to use the rapidly expanding Christian religion, with its great potentials, to strengthen the Roman Empire. (163)

Giving Christ equality with God legitimized the Christians; the motive for establishing Jesus' deity was pure power politics! Jesus never claimed equality with God, whose mission he was sent to fulfill.

A third important teaching was that Jesus came to the Hebrews only. He was sent by the Starmen to gather the "lost sheep of Israel" and to reestablish the righteous remnant of Israel as servants of the Lord or to break the covenant between God and his people - forever.

"I was sent only to the lost sheep of the house of Israel." (Matt. 15:24)

These twelve [apostles] Jesus sent forth, and commanded them saying, "Go not unto the way of the Gentiles...but go rather to the lost sheep of the house of Israel." (Matt. 10:5-6)

It was, again, not Jesus but later editors and theologians who contended that Jesus was concerned with all mankind. In fact, one instance in particular sheds light on Jesus' feelings toward his wasting time with the gentiles when, in a rare moment of candor, Jesus tells a gentile woman begging for help for her sick daughter, "It is not fair to take the children's bread and throw it to the dogs." (Matt. 15:25) It is perhaps not that Jesus did not want to help, for he was a compassionate man, but that his resources and time were severely limited, and his mission was strictly for his people.

A fourth important teaching, made quite explicit, is that Jesus was the product of another world:

"You are from below, I am from above; you are of this world, I am not of this world." (Jn. 8:23)

Just before Jesus was to be delivered up to be crucified, he spoke of this world to which he would be removed:

"Where I am going you cannot come."...Simon Peter said to him, "Lord, where are you going?" Jesus answered, "Where I am going you cannot follow me now; but you shall follow afterward." (Jn. 13:33, 36)

What kind of place was this that Jesus spoke of? When pressed further, Jesus replied obliquely:

"Let not your hearts be troubled; believe in God, believe also in me. In my Father's house are many rooms; if it were not so, would I have told you that I go to prepare a place for you?" (Jn. 14:1-2)

Jesus must have been questioned about this strange place and about his peculiar ways more than a few times, for such teachings could only have fueled the disciples' curiosity. Nevertheless, such concepts as life on other planets, spaceships and transport through space would not only have been useless for the mission, but likely incomprehensible to the disciples as well. As once decreed by God in Isaiah;

"For as the heavens are higher than the earth, so are my ways higher than your ways, and my thoughts higher than your thoughts." (Is. 55:9)

Jesus' actions and teachings reflect his sophistication and, despite the current necessity for his metaphorical double-talk, he promises that soon he will reveal more:

"I have said this to you in figures; the hour is coming when I will tell you plainly of the Father." (Jn. 16:25)

A final key teaching for the Starmen Theory is Jesus' pronouncement that his kingdom was at hand and that some would still be alive when it arrived:

"Truly I say unto you, this generation will not pass away till all has taken place." (Jesus to disciples; Luke 21:32)

"Children, it is the last hour; and as you have heard that the antichrist is coming, so how many antichrists have come; therefore we know it is the last hour." (I John 2:18)

"Truly, I say unto you, there are some standing who will not taste death before they see the kingdom of God come with power." (Mk. 9:1)

Thus, whatever the plan, the end was to occur in the lifetimes of the original apostles, and the establishment of the new kingdom would be witnessed by them also. But we know that this kingdom did not arrive during their lifetimes, and it has not yet arisen some two thousand years later. The promise made by Jesus was never fulfilled, for instead of ascending to the throne of Israel and restoring her to her former glory, he was rejected and crucified by the very ones he sought to save. Before we elaborate on this, however, let us talk about some of the miracles attributed to Jesus.

There are many interesting and extraordinary events in Jesus' life and ministry. Most of the miracles during this period concerned his healing of various ailments and deformities, from curing lepers to reversing impotency, from exorcising demons to raising the dead, and from straightening a withered hand to restoring sight to the blind. We briefly discussed the possibilities behind some forms of healing through faith in the healer and the mind's control of the body's natural life forces in a previous section. Whether or not Jesus actually did all of the things he is credited with cannot be determined. Perhaps he had help from the Starmen and an understanding of medicine which we have yet to discover; perhaps they were staged events; perhaps they were myths.

A few other "miracles" were performed, such as changing water into wine and feeding multitudes of people from only a few meager supplies. These likewise must be acknowleged without comment for their circumstances and context shed no light on their cause. But there are a couple of interesting events which are outside of these other types of "miracles" and seem to represent phenomena already discussed as relating to the Starmen vehicles and mission.

The first deals with a possible encounter with a vehicle not unlike that experienced by Elijah or Moses except that the tumult caused by the vehicle was not on land, but over the water. After Jesus had healed many sick people at Capernaum, he went to the shore of the Sea of Galilee to escape the crowds, who kept pressing him to heal more:

And when he got into the boat, his disciples followed him. And behold, there arose a great storm on the sea, so that the boat was being swamped by the waves; but he was asleep. And they went and woke him, saying, "Save us, Lord; we are perishing." And he said to them, "Why are you afraid, O men of little faith?" Then he rose and rebuked the winds and the sea; and there was a great calm. And the men marveled saying, "What sort of man is this, that even winds and sea obey him?" (Matt. 8:23-27)

Are we witnessing a great triumph over the elements, or was this "storm" caused by a cloud-enveloped vehicle hovering over the small boat? Perhaps Jesus realized what was happening and waved the craft away. Consider again the opening lines of Ezekiel for comparison:

As I looked, behold, a stormy wind came out of the north, and a great cloud, with brightness round about it...(Ezek. 1:4)

Consider also the effects reported in I Kings:

And behold, the Lord passed by, and a great and strong wind rent the mountains, and broke in pieces the rocks before the Lord...and after the wind an earthquake... (I Kings 19:11)

What other "storm" is there that a man can simply "rebuke" and wave off, as did Jesus? In any case, they crossed safely and continued on their way as if nothing had happened.

A second incident worth noting is the report that Jesus walked on water. Again we have wind in association with the boat on the water:

Immediately he made his disciples get into the boat and go before him to the other side, to Bethsaida, while he dismissed the crowd. And after he had taken leave of them, he went into the hills to pray. And when evening came, the boat was out on the sea, and he was alone on the land. And he saw that they were distressed in rowing, for the wind was against them. And about the fourth watch of the night he came to them, walking on the sea. He meant to pass by them, but when they saw him walking on the sea they thought it was a ghost, and they cried out; for they all saw him, and were terrified. But immediately he spoke to them and said, "Take heart, it is I; have no fear." And he got into the boat with them and the wind ceased. (Mk. 6:45-51)

An orchestrated event? A miracle? Remember various passages quoted earlier of the events produced by the "spirit" of the Lord and in which Enoch, Elijah and Ezekiel were lifted up and apparently carried through the air as if by magic. If this was an orchestrated event, Jesus obviously had plenty of time to coordinate with the Starmen while "praying" alone in the hills. Perhaps he needed to get over to his disciples, who he thought were on the other side by the fourth watch, and so was expediently carried by the "spirit" of the Lord over the sea. But when his disciples spotted him crossing over the surface, seemingly walking on the water, they cried out to him. They were astounded, and they thought they were seeing things. When Jesus realized that his disciples had seen him, he directed the vehicle transporting him to take him to them and was released into the boat. Then the vehicle above departed, taking its wind with it.

Preposterous? More so than walking on water? If the Starmen were present, it would be logical to assume that they would have left some marks, inexplicable to the primitives, which eventually became our "miracles." It should not be forgotten that such demonstrations were the way God revealed himself or his proxies to numerous Old Testament hosts as well, and thus these two incidents fit nicely into the thesis of The Order, regardless of the original motives behind them.

The question of when Jesus or the Starmen realized that their final attempt to retrieve their kin was in vain will perhaps always remain a mystery. It was a calculated risk, and Jesus seems to have understood this from the beginning. The appeal of his message to the poor masses, who had little or nothing in the way of worldly goods, could have been predicted. Without, however, the defection of a number of powerful political-religious sectarians, his work was also predictably to be found contrary to their interests, and his mission therefore ultimately a failure. The price, as we know, was a cruel death by crucifixion. But herein lies another question: would the Starmen have allowed their hybrid kin to be martyred?

It seems inconceivable, even under the extraordinary circumstances, that the Starmen would eventually allow their proxy to suffer death at the hands of the wayward Hebrews or that Jesus could offer himself to be killed. In fact, Schonfield believes that despite the social, religious and political passions and Jesus' own conviction that he was the awaited Messiah, Jesus only wanted to *stage* his death so that he could actually appear to have been physically resurrected from the dead and then continue on with his ministry. But instead the plan failed, and he was martyred.

Did the Starmen have a contingency plan to allow Jesus to appear to be killed if he failed to rally his people? The events of the Bible seem to bear this out in two ways: First, we know that Jesus originally came to gather his "lost sheep," but when he is rejected, he made plans to become a martyr for his cause, presumably to lend force to the momentum already built. Second, we know from the scriptures that Jesus was taken *alive* back into heaven. It must therefore be assumed that the decision to stage Jesus' death was made at some point well into his ministry.

The Starman Intervention Theory thus takes Schonfield's idea one step further and proposes that Jesus acted under the direction of, and in concert with, the Starmen to stage his death and resurrection for his people and was successfully taken alive back into space after the event.

Once the decision was made to go ahead with the risky new plan, Jesus had to convince his closest aides of the gravity of the situation and of the new agenda. In order to win the unwavering support of his closest lieutenants—Peter, James and John—

Jesus had to reveal to them that he was not acting alone and that the mission carried the weight of a superior order which required total allegiance. The Gospel of Matthew tells us how this was achieved:

> *...after six days Jesus...led them up a high mountain apart. And he was transfigured before them...and his garments became white as light. And behold, there appeared to them Moses and Elijah, talking with him. [Then] a bright cloud overshadowed them and a voice from the cloud said, "This is my beloved Son, with whom I am well pleased; listen to him."* (Matt. 17:1-9)

In a natural reaction to the awesome sight, the disciples immediately fell on their faces in fear. Then Jesus bade them to stand up and charged them to tell no one until he had died and had been raised from the dead. Whether the three apostles really saw Elijah and Moses or just projected their immediate impressions onto two unknown Starmen is not made clear. Other characteristics of this event provide us with solid continuity to earlier recorded events.

For example we note again the "bright cloud" and the voice emanating from it, which marks the third such recorded contact in the New Testament and is quite consistent with the earlier Old Testament visitations. Likewise this event seems to have struck fear into these poor primitives, just as with the shepherds, with Ezekiel, with Elijah and with all who came into contact with such vehicles in the Bible accounts.

With this simple act, the superior force had manifested itself to the apostles, and their minds were then understandably firmly set to help Jesus in any way necessary to achieve his goals even if they did not fully understand the purpose of the plan.

The final event of significance to the Starman Theory is the resurrection of Christ. Each of the Gospels tells of the resurrection, but each version varies to a degree, mostly through the inclusion or exclusion of material, depending upon what the authors were trying to convey about Jesus.

The Gospel of Matthew provides us with the most detailed account of the event. It is a fascinating story which begins with a man named "Joseph of Arimathea," who went to Pontius Pilate (the Roman perfect of Judea who had sentenced Jesus), after Christ's supposed death and asked for the body of his teacher. Pilate obligingly ordered the body released to Joseph, who then wrapped it in a clean linen and placed it in a tomb hewn out of solid rock, a tomb, we are told, which Joseph had built for himself. We are also told that the door to the tomb was a great stone - probably hewn in the shape of disk - which was rolled down a grooved track to seal the entrance.

Meanwhile the Sadducees and the Pharisees, though greatly relieved at Jesus' elimination, were nevertheless uneasy that more trouble could still be stirred up among the people if Jesus' body were stolen from the tomb by one of his followers, who might then proclaim that Jesus had arisen from the dead. To prevent this, Pilate granted permission to station Roman guards outside the tomb for three days, until the danger of Jesus' prophesized resurrection had passed. When the soldiers arrived, the tomb was sealed and then kept under constant guard. What happened next is perhaps best told by the scriptures themselves:

> *Now after the sabbath, toward the dawn of the first day of the week, Mary Magdalene and the other Mary went to see the sepulchre. And behold, there was a great earthquake; for an angel of the Lord descended from heaven and came and rolled back the stone, and sat upon it. His appearance was like lightning, and his raiment white as snow. And for fear of him the guards trembled and became like dead men. But the angel said to the women, "Do not be afraid; for I know that you seek Jesus who was crucified. He is not here; for he has risen, as he said."* (Matt. 28:1-6)

The Gospel of Luke adds that Jesus was taken alive into heaven, for an angel asks the two women, "Why do you seek the living among the dead?" (Luke 24:5) And we are told he was carried away in a cloud:

> *...as they were looking on, he was lifted up, and a cloud took him out of their sight. And while they were gazing into heaven as he went, behold two men stood beside them in white robes and said, "Men of Galilee, why do you stand looking into heaven? This Jesus, who was taken up from you into heaven, will come in the same way as you saw him go into heaven."* (Acts 1:9-11)

What one account calls an "angel of the Lord," another calls a "cloud." Still, both descriptions are by now quite familiar. The "angel" in Matthew was a very bright light, "like lightning," which descended over the tomb, causing a "great earthquake" and striking terror into the guards. From this vehicle came the voice of the pilot who spoke, as at Jesus' baptism, through a loudspeaker to the people in the vicinity.

Apparently two Starmen, described *not* as angels, but as "men in white robes," remained behind for a time to speak to the people or to carry out other activities not mentioned. Even with different witnesses from diverse backgrounds, locations and times, a remarkable degree of continuity is evident in all accounts!

Thus the scriptures assure us that Jesus was a Starman hybrid who "for a little while was made lower than the angels" (Heb. 2:9) to try to "write" the laws of God on the hearts of the Hebrews that they might be turned to God through their own wills instead of through the old method of coercion by fear. Failing to arouse sufficient momentum to overcome existing political and religious powers, the Starmen opted to make Jesus a martyr to instigate a groundswell of grassroots support by the masses.

And here we have another turning point in the philosophy of the Starmen. Sometime between the "resurrection" of Jesus and the formation of the early church, the Starmen, in a critical realignment of their thinking, decided to broaden their base of support to *include* the gentiles in a movement to learn of Jesus' teachings. This was a remarkable turn if true. Why would the Starmen opt to include those who never shared in the Hebrews' special relationship to God?

The answer to this question is quite simple and yet quite profound. First we recall that in II Peter 3:8 the message is that "with the Lord one day is a thousand years, and a thousand years as one day." In other words, time by our reckoning is of little significance, if any, to the Starmen in their space travels from one galaxy to the next. With the *battle* over the hearts and minds of their Hebrew kin lost, could they hope to win the *war* by passing Jesus' precepts on to all mankind? What the Hebrews for the most part rejected might be accepted by sufficient numbers of gentiles to perpetuate them into the future! And the mechanisms of this perpetuation? The Church!

Did Jesus teach acceptance of his teachings on faith because he knew that much of what he taught would not be fully understood for many centuries? "Truly I say unto you, unless you turn and become like children, you will never enter the kingdom of heaven." (Matt. 18:3)

Perhaps the Starmen understood that the unadulterated perpetuation of their precepts into the future would depend upon a method of knowledge preservation fostered by the faith of ignorant minds through a continual passing down of that which they would not, indeed could not, fully understand for generations. As children must accept many things on faith until they mature, mankind was also tasked to accept The Order on faith until we had matured both technologically and socially. Why would the Starmen want to pass these precepts on into the future? So that they could be properly understood in their own time.

We are now becoming as angels ourselves. Jesus knew that if man did not destroy himself, he would one day become as an angel and move into space, as we are now doing. And when his ability to reach the inhabited planets of other galaxies is attained, he will be initiated into the intergalactic order of the heavens. He will then be an angel under the laws of the Lord of a federation of worlds. And if he cannot meet the requirements of these laws, he will have to be destroyed as a menance to The Order:

"But one day the Lord will come like a thief, and the heavens will pass away with a loud noise, and the elements will be dissolved with fire, and the earth and the works that are upon it will be burned up." (II Peter 3:10)

Although the original Christians and thousands of generations beyond them would never understand the true and complete significance of Jesus' precepts, the perpetuation of those ideas would mold and prepare future generations of mankind to move into space with moral authority. And should they come upon other less sophisticated races, they might be kind and careful not to corrupt the delicate order which extends throughout the universe, as the watchers had done on Earth.

The *Book of Enoch* is a non-canonical book which was written and assembled from 165 B.C. to about 64 B.C. It was not included in the Bible for various political reasons even though it was widely read and enjoyed by the Essenes and likely Jesus himself. It, like Daniel and Revelations, is an apocalyptic book of predictions of possible future events. In it is an opening passage, apparently written by Enoch himself:

Enoch a righteous man, whose eyes were opened by God, saw the vision of the Holy One in the heavens, [which] the angels showed me, and from them I heard everything, and from them I understood as I saw, not for this generation, but for a remote one which is to come. (Enoch 1:2)

There remains yet a key element to discuss at this point: the transfer of Jesus' teachings to the gentiles and the establishment of the Church mechanism which would perpetuate them.

129. Werner Keller, *The Bible as History,* Bantam Books, New York, 1956, p. 66.

130. *Ibid.,* p. 118.

131. Raymond W. Drake, *Gods and Spacemen of the Ancient Past,* Signet Books, New York, 1974, p. 201.

132. *Ibid.,* Book IV, Chapter VIII, p. 48.

133. Keller, *op. cit.,* pp. 171-2.

134. Miller, *op. cit.,* p. 648.

135. *Ibid.,* p. 87.

136. *Ibid.,* pp. 692-4.

137. Josef F. Blumrich, *The Spaceships of Ezekiel,* Bantam Books, New York, 1974.

138. Miller, *op. cit.,* p. 181.

139. Blumrich, *op. cit.,* p. 114.

140. *Ibid.,* p. 13.

141. *Ibid.,* p. 131.

142. *Time,* April 20, 1981, advertisement for.

143. Blumrich, *op. cit.,* p. 23.

144. Charles H. Patterson, *Cliff's Notes on the Old Testament,* C.K. Hillegass Pub., Neb., 1965, pp. 47-8.

145. "Jesus," *Merit Student Encyclopedia,* 1978, p. 164.

146. Miller, *op. cit.,* p. 428.

147. *Ibid.,* p. 419.

148. *Ibid.*

149. Hugh J. Schonfield, *The Passover Plot,* Bantam Books, Inc., New York, 1965, p. 15.

150. *Ibid.,* p. 17.

151. Rev. Dr. Charles Francis Potter, *The Lost Years of Jesus Revealed,* Fawcett Publications, CN, 1962, p. 23.

152. *Ibid.,* p. 19.

153. Miller, *op. cit.,* p. 173.

154. Erich Von Daniken, *Miracles of the Gods,* Dell Publishing Co., Inc., New York, 1975, p. 75.

155. Potter, *op. cit.,* p. 16.

156. *Ibid.*

157. James L. Roberts, Ph.D., *Cliff's Notes on the New Testament,* C.K. Hillegass Pub., 1965, p. 17.

158. Miller, *op. cit.,* p. 836.

159. Roberts, *loc. cit.*

160. Platt, (Lost Books), *op. cit.,* Protovangelon; IX:15.

161. Lee Roddy and Charles E. Sellier, Jr., *In Search of Historic Jesus,* Bantam Books, New York, 1979, pp. 31-2.

162. Schonfield, *op. cit.,* p. 61.

163. Von Daniken, *op. cit.,* p. 57.

Ezekiel 3:12-15

Luke 2:8-15

Chapter XI
THE NEW ORDER

Is there anything of which it may be said, "See, this is new"? It has already been in ancient times before us.

— Eccles. 1:10

The Church in the beginning was made up exclusively of the Hebrew followers of Jesus, and that is the way it continued for several decades after Jesus' "resurrection." Unlike the Sadducees and the Pharisees, these Hebrews believed that Jesus *was* the awaited Messiah and, as such, was only the latest occurrence in a long line of continuous works by God, the same God of Abraham and Moses. Their head figure was the apostle Peter, the "rock" upon which Jesus once proclaimed that he would build his church.

The Hebrews still believed in God's selection of their race (or the righteous remnant thereof) as the exclusive instruments of His will and thus rejected all foreign gods and viewed most pagan practices with disdain. They still held that it was their divinely appointed destiny to be a priestly race unto all mankind and set themselves apart from all others in their Greco-Roman environment. This led to what Chadwick called "some degree of vulgar distrust and at times hostile prejudice" against them. (164) Their revolutionary ideas and cohesive nature made them natural targets for slander, scapegoating and other persecutions.

Centuries of foreign domination and exile, combined with the poor economy of Palestine, caused the Hebrews to scatter all over the Mediterranean in what has become known as the "Dispersion." As a result Hebrew colonies spread from Cadiz to the Crimea, where they formed rather exclusive, compact social communities in most cases.

In some instances, however, it appears that these tightly knit societies-within-societies began to open their doors not only to some of the foreign manners, but to those interested in their ways as well. Chadwick reveals that although such practices as baptism and circumcision were scorned by Greeks and Romans, "many Gentiles were attracted by monotheism, by the purity of Jewish morality, and by the antiquity (if not by the style) of their sacred books." (165) According to Chadwick:

...Judaism stood for chastity and stable family life; and among themselves the Jews practiced works of charity, visiting the sick, caring for the dead, showing hospitality to strangers, giving alms for the poor. (166)

At first the gentiles who wanted to join the movement were subject to much of the Mosaic law which was binding upon the Jews, including the rites of circumcision and baptism. There was some contention in the early Church about such matters. While the hellenized Jews of the Dispersion were content to allow gentiles to forego circumcision and other Jewish ceremonial practices, the Palestinian Jews wanted stricter adherence to the Laws of Moses as a prerequisite for salvation. As described earlier, the Jews at the time of Christ were of diverse ideologies, and the assimilation of the gentiles by a few of the more liberal sects was a natural evolutionary step in the growth of the Church.

Meanwhile, as the gentile membership in the Church grew, it spread from Jerusalem to the surrounding countryside of Judea and further, to Damascus and even Antioch, the capital of Syria. The movement came to be known as "the Way," and its followers "Christians." (The Jewish term for them was "Nazarenes"). Such rapid expansion was not without repercussion however.

The Jewish establishment, the mainline Jews who refused to compromise on the subject of gentile entrance into the Church, felt such anxiety over the rapid rise of this new cult that a counter-movement was intitiated against them. These establishment Jews, who maintained a conservative profile, pursued a policy of compromise and even collaboration with the occupying Roman government and thus held the upper hand. Leading this movement was Saul of Tarsis, a Cilician Jew and former pupil of the Pharisee Rabbi Gamaliel at Jerusalem.

Saul was born in what is now Turkey in 4 A.D. The first part of his life he spent persecuting what he considered the deviant Hebrew sects, including the new branch of "Christians," because he was convinced of the literal and dogmatic perfection of the Mosaic Law. It was Saul who had set out on the road to Damascus one day on such a mission of persecution when he was confronted by a "vision" of Jesus, the "resurrected Christ" only days after his crucifixion, and miraculously converted from persecutor to apostle. The *Book of Acts* describes the event which so altered Saul's convictions:

Now as [Saul] journeyed he approached Damascus, and suddenly a light from heaven flashed about him. And he fell to the ground and heard a voice saying to him, "Saul, Saul, why do you persecute me?" And he said, "Who are you, Lord?" And he said, "I am Jesus, whom you are persecuting; but rise and enter the city and you will be told what to do." The men who were traveling with him stood speechless, hearing the voice but seeing no one. Saul arose from the ground; and when his eyes were opened, he could see nothing; so they led him by hand and brought him into Damascus. (Acts 9:1-7)

By any measure this event must have been quite powerful to turn an educated man of Saul's conviction 180 degrees in his philosophy toward the Christians he once had persecuted. His peers were shocked at his defection. Downing compares the turnabout to something like Hitler suddenly realigning with the allied army during World War II! (167) Note how this "vision" too is consistent with previous Starmen contacts.

While the various accounts of this event recorded in the New Testament do not all jibe exactly, enough elements exist in the preceeding Acts passage to verify the Starmen's presence. The bright light of the vehicle which came down from "heaven" also produced the voice from the unseen speaker (who identified himself as Jesus) and Saul's blinding, as at Sodom, provide the requisite continunity. Upon reaching Damascus, Saul's sight was restored, and he set about to learn the ways of the Christians. (See also; Acts 22:6 & 26:12-18)

Saul's gentile name was "Paul," a name that has remained prominent throughout history. After his conversion, he became the most dynamic and significant individual in the formation of the early Church. So influential was Paul to the movement that he is given nearly equal rating to Jesus and the title of "founder" of the Christian religion! In his book *The 100,* M.H. Hart states:

Christianity, unlike Islam, was not founded by a single person but by two people - Jesus and St. Paul - and the principal credit for its development must therefore be apportioned between these two figures. (168)

Paul is thus given the main credit for founding a church from the precepts which Christ taught to the Jews even though he never saw Christ while he was alive on earth! The situation being as it was, with the recent infiltration of the gentiles, gave Paul reason to initially question the gentile involvement in something that was historically of Hebrew concern. His conclusion was, of course, that the gentiles were to be an integral part of the new movement, and indeed they eventually replaced the Hebrews as the dominant population of the religion. Paul so thoroughly defended their inclusion, in fact, that he became known as the "apostle of the gentiles."

Paul traveled extensively throughout the Roman Empire to help set up churches and to settle disputes over doctrinal matters among the followers. He is credited with writing at least fourteen of the twenty-seven books in the New Testament, making him the single most significant writer and thinker in Christian history. It is he who lent to Jesus his "divine" character, with its spiritual connotation, and he who broke down the major barriers in the Jewish religion which made it possible for the gentiles to share Jesus' teaching and hope. Again quoting from Hart:

Paul more than any other man, was responsible for the transformation of Christianity from a Jewish sect into a world religion. His central ideas of the divinity of Christ and of justification by faith alone have remained basic to the Christian thought throughout all intervening centuries. (169)

Paul was the gentile's shepherd. He appealed to the poor and the downtrodden. It was his interpretation that much of the Mosiac law—such as circumcision, ritual and dietary practices—applied mainly to the Jews. It was he who perceived that all were equal in the eyes of God, and his theology cut across both sexual and racial lines to forge an integrated, compact movement which respected the rights of all people.

According to Hart, Paul's overwhelming influence resulted from three major factors: first was Paul's, and to a lesser extent Peter's, aggressive missionary work throughout the Roman empire; second was the fact that he made major contributions to the writings of the New Testament and wrote other essays widely read at the time and, which won him high visibility and credibility; and third was the tremendous impact his theology had on subsequent biblical writers and thinkers. (170)

If Christianity's widespread appeal prompted its rapid expansion, adversity served to solidify it. Persecution existed sporadically from the beginning. It was sometimes cruel and unwarranted, but those who died and became martyrs only served to bind the Christians together, to stiffen their resolve, and to advertise their religion to others who might become converts. So rapidly did this cult grow that Rome itself, in a dramatic reversal of philosphy, was forced to accept it into the circle of respectable religions during the reign of the Emperor Constantine in 313 A.D. By that time some twenty million people, almost one-quarter of the entire Roman empire, had converted to Christianity.

This official edict of tolerance was more a political move than a divine victory, however, for Constantine wisely chose this course to help expand his empire. When faced with the opportunity to overthrow Licinius, his arch rival of the East, Constantine called upon the Christians and their God, thereby rallying enough might to overthrow his adversary and become emperor of the entire empire. Thereafter Christians enjoyed a much more respectable and comfortable environment. But it was not until the year 380 that Theodosius the Great raised Christianity from among the other religions to the sole religion of the empire. To underline the significance of his sweeping mandate, Theodosius declared all pagan religions a crime against the state! By 529 the conversion of the empire was complete. Paul's Christianity had grown so explosively, even in the face of occasional state sanctioned persecution, that an entire empire was thus converted within the span of five centuries.

The exact nature of this growth and expansion is a history complex and tumultuous, a sequence of events perhaps not altogether as the Starmen would have planned but easily recognizable as the work of men, inspired or not. As stated earlier, Christianity has established itself among the world's greatest powers, preeminent among religions if not the most significant influence on the ethics of mankind. We must ask then, has the mission of the Starmen been a success after all?

* * *

Thus the birth of Christianity has given way to a flood of diverse offshoots, splinter groups and theme variations, most of which are straying far from the original Hebrew precepts. Today, as never before, this diversity is increasing with an amazing capacity for endless permutations on the original theme. So even if the Christian religion is gaining new adherents, it is not clear that they will be educated in the manner intended by the Starmen. It is as if the gentiles now are corrupting the original beliefs as the Hebrews were once accused of doing over and over in the Old Testament. Now, as we stand at the threshold of our move out into space, do we meet the requirements for acceptance into the Starmen's domain?

It seems reasonable that even as Christianity changes and diversifies itself into oblivion, man's discovery of the true nature of the order of things recorded in the Bible will allow sufficient numbers of us to glean the ethical message from the story and combine it with secular ethical standards to achieve the move into space with the requisite morality.

Some have said that without the God of the Bible, organized religion and all its trappings—even though it is corrupt in many ways—is still better than the secular alternatives or nothing at all. At least the so-called "fear factor" causes people to try to behave correctly, and that doesn't seem like such a bad thing to them. But that may not be the case at all.

For one thing, just knowing the proper way to interpret the stories of the Bible, being able to logically explain the miracles and the God and stories of the Bible, would certainly prevent most people from worshipping the Starmen as we do now. Most people would, it might be hoped, turn to secular ethical systems, such as the ones taught by the great philosphers, or even espouse tenets of the eastern religions which, to a remarkable degree, promote the appreciation of the natural order of things and man's role in it.

But one thing is clear: to continue worshipping the God of the Bible the way He is now being worshipped, whether he is a Starman or not, will do little to unify mankind for the trend in Christianity appears to be toward endless modification, not stabilization. But our ability to unite and to move out into space together, with moral authority, seems to be contingent on shedding the cocoon of Christianity and emerging into a completely new perception of our existence and our relationship to God.

Is there a God? It is by now obvious that the author's belief is predicated upon an observable natural order, whether mechanistic or dynamic. It is a form of Deism which advocates natural religion based on reason rather than revelation and emphasizes nature's order and intelligibility. Within this order the cumulative knowledge of man is increasing. Does this unfolding knowledge indicate a directive force in nature, or are the laws themselves fixed with man awakening to discover that he is the god which has arisen from the primordial soup?

At this time I remain agnostic, knowing I lack the facts necessary to form concrete opinions about whether or not God is a conscious entity. But I have no fear, for what I see is a comprehensible system even if its origins and workings are yet a mystery. My goal is to improve myself and to study how to be a good and ethical being. Knowing that the Lord of the Bible was a Starman has allowed me to release my childhood fear of Hell. Without that fear, I don't need the Church's absolute code of conduct, which produces the guilt for not having lived up to unattainable standards. I now behave as well, or better, than I did before without all of the Christian trappings. My desire to do what is right is based upon a few simple principles, such as: do unto others as you would have them do unto you; act so as to produce the greatest good for the greatest number; and act with a conscience.

When I die, all will not have been in vain for I will have left a legacy on Earth, and that is my reconciliation with death. This legacy which we leave behind after we die is threefold. First, our bodies return to the earth to replenish it. We give up only what we borrowed. Second, our deeds on Earth are preserved in the memories of others and in tangible contributions we made on Earth while alive. Finally, our children, as living extensions of ourselves, will carry on. Our purpose in life, therefore, is to contribute to the survival of the human race by doing our individual best within the laws of society and the laws of nature.

If we follow these natural laws so evident to the ancients and so often ignored by modern man, then we shall perhaps one day actually meet people of other worlds and even join them in an intergalactic society as descendants of the same God. Until then we are undergoing a trial period of discovering the path to survival.

Prudence, moderation, courage, knowledge, integrity and justice are the virtues which will favor our perpetuation. The appreciation of knowledge is to be emphasized for from the proverbial "tree of knowledge" issues forth wisdom. Critical thinking then reveals the ethical standards by which we will shape our society in the pursuit of utopia - or heaven, if you will. All life derives from an understanding of The Order. Repudiation of its laws will lead to chaos, a living hell on Earth, and even the extinction of our planet. It is only when we realize the importance of these interrelationships and make the next meeting with our galactic neighbors that we will know if the mission of the Starmen was a success or a failure.

164. Henry Chadwick, *The Early Church,* Penguin Books, Ltd., New York, 1967, p. 9.

165. *Ibid.,* p. 11.

166. *Ibid.*

167. Downing, *op. cit.,* p. 133.

168. Michael H. Hart, *The 100,* Hart Publishing Co., New York, 1978, p. 47.

169. *Ibid.,* p. 64.

170. *Ibid.,* p. 62.